TUTOR DELIVERY PACK

MATHEMATICS

— GCSE —

FOUNDATION

 Pearson

 Tutora

CONTENTS

CONTENTS

HOW TO USE THIS PACK

The *Tutors' Guild* Maths Tutor Delivery Pack gives you all of the tools you need to deliver effective Maths lessons to AQA GCSE (9–1) students who are sitting the Foundation paper. Everything in this pack is available for you to download as an editable file. This means that every lesson can be edited to suit the needs of your student, but also that you can print off each resource as many times as you need.

LESSONS

There are 38 one-hour, six-page lessons in this Tutor Delivery Pack. Most tutors working for a full year will have around 38 lessons with a student. If you have less contact time, you can choose which lessons are most important to the student and build your own course, using the customisable digital version of this pack. Each lesson is standalone and can be taught independently from those preceding it.

If you have more than 38 lessons together, or your lessons are longer than one hour, you can incorporate assessment from the accompanying Tutor Assessment Pack (ISBN: 9781292195551). There is an *end-of-topic test* for every lesson in this pack, as well as *checkpoint challenges* and a *practice paper*. All of the papers can also be given as homework, used as diagnostic tests or incorporated into revision.

LESSON PLANS

The first page of each lesson is your *lesson plan*. It is designed specifically for tutors and is intended to guide you through a one-hour session in either a one-to-one or small group setting. It is not designed to be student-facing.

LEARNING OBJECTIVES AND SPECIFICATION LINKS

At the top of each lesson plan, you will find two lists. The first – *learning objectives* – is a list of your aims for the lesson. The learning objectives will be informed by the specification but may have been rephrased to make sure they are accessible to and useful for everyone. You can discuss these with the student or use them for your own reference when tracking progress. The second list – *specification links* – shows you where in the specification you can find the objectives relevant to the lesson. You can find out more about the specification on pages 7–10.

ACTIVITIES

The first five minutes of your lesson should be spent reviewing the previous week's homework. You should not mark the homework during contact time: instead, use the time to talk through what the student learned and enjoyed, and any difficulties they encountered.

The final five minutes should be used to set homework for the forthcoming week. There are three ways to do this: using the *end-of-lesson report* on page 16; orally with a parent or guardian; or simply using the *homework activity sheet* on the fifth page of each lesson.

In each lesson plan, you will find four types of activities:
- *starter activities* are 5–10 minutes each and provide an introduction to the topic
- *main activities* are up to 40 minutes long and are more involved, focussing on the main objectives of the lesson
- *plenary activities* are 5–10 minutes each, require little to no writing and recap the main learning points or prepare for the homework
- *homework activities* can be up to an hour long and put learning into practice.

In the lesson plan, you will find a page reference (where the activity is paper-based), a suggested timeframe and teaching notes for each activity. The teaching notes will help to guide you in delivering the activity and will also advise you on any common misconceptions associated with the topic.

HOW TO USE THIS PACK

SUPPORT AND EXTENSION IDEAS

This pack is aimed at students who are targeted grades 3–5, but every student is different: some will struggle with activities that others working at the same level find straightforward. In these sections, you will find ideas for providing some differentiation throughout the activities.

PROGRESS AND OBSERVATIONS

This section is left blank for you to use as appropriate. You can then use the notes you make to inform assessment and future lessons, as well as to inform *progress reports* to parents.

ACTIVITY SHEETS

There are four student-facing *activity sheets* for each lesson: one for the starter activities; two for main activities and one for the homework activity. On each sheet, you'll find activity-specific lesson objectives, an equipment list and a suggested timeframe. All activities are phrased for one-to-one tutoring but are equally as appropriate for small group settings. If you have a small group and the task asks you to work in pairs or challenge each other, ask the students to pair up while you observe and offer advice as necessary. Where appropriate, answers can be found on the sixth page of the lesson.

DIAGNOSTICS

The first lesson in this pack is a diagnostic lesson, designed to help you find out more about your student: their likes and dislikes; strengths and weaknesses; and personality traits. As well as the diagnostic lesson, the *needs analysis* section (pages 13–14) allows you, the student and the student's parents to investigate together which areas of the subject will need greater focus. Together, these sections will help you deliver the most effective, best value tuition.

PROGRESS REPORT

This can be used to inform parents or for your own planning as frequently or infrequently as is useful for you. Spend some time discussing the statements on the report with the student. Be prepared, though – some students will tell you there isn't anything that they enjoy about the subject!

END-OF-LESSON REPORT

Parent participation will vary greatly. The *end-of-lesson* report is useful for efficiently feeding back to parents who prefer an update after each lesson. There is space to review completed homework and achievements in the lesson, as well as space for the student to explain how confident they feel after the lesson. Finally, there is a section on what steps, including homework, the parent and student can take to consolidate learning or prepare for the following week. The *end-of-lesson* report may also be useful for communicating with some parents who speak English as a second language, as written information may be easier to follow.

CERTIFICATES

In the digital version of this pack, you will find two customisable certificates. These can be edited to celebrate achievements of any size.

PEARSON PROGRESSION SCALE

 Each question in this pack features a Step icon that indicates the level of challenge aligned to the Pearson Progression Map and Scale. To find out more about the Progression Scale for Maths and to see how it relates to indicative GCSE 9–1 grades go to www.pearsonschools.co.uk/ProgressionServices

MATHS
— FOUNDATION —

INFORMATION FOR PARENTS AND GUARDIANS

INTRODUCTION

Your child's tutor will often make use of resources from the *Tutors' Guild* series. These resources have been written especially for the new GCSE (9–1) qualifications and are tailored to the AQA Mathematics specification. The tutor will use their expert knowledge and judgement to assess the student's current needs. This will allow them to target areas for improvement, build confidence levels and develop skills as quickly as possible to ensure the best chance of success.

Just as a classroom teacher might do, the tutor will use lesson plans and activities designed to prepare the student for the new (9–1) GCSEs. Each set of resources has been designed by experts in GCSE (9–1) Mathematics and reviewed by tutors to ensure it offers great quality, effective and engaging teaching. All *Tutors' Guild* resources are flexible and fully adaptable, so you can be confident that the tuition the student receives is tailored to his or her needs.

GETTING STARTED

Before tuition can begin, the tutor will need to know more about your motives for employing them in order to set clear, achievable goals. They will also try to learn more about the student to ensure lessons are as useful and engaging as possible.

To gather this information, the tutor will work through the *needs analysis* pages of this pack with you. It shouldn't take too long, but it will really maximise the value of the tuition time you pay for. You could also take this opportunity to discuss with the tutor any questions or concerns you may have.

LESSONS AND HOMEWORK

Each lesson will have the same structure: there will be a starter, which is a quick introduction to the topic; some main activities, which will look at the topic in greater detail; and a plenary activity, which will be used to round off the topic. Throughout the year, the student will become increasingly confident with the content of the specification, but will also improve his or her numerical fluency, problem solving skills and multiplicative reasoning through a carefully balanced range of activities.

At the end of each lesson, your tutor will set some homework, which should take no longer than an hour to complete. If you don't want the tutor to set homework, please let them know. If you are happy for homework to be given, the tutor will either discuss the homework task with you at the end of the lesson or give you an end-of-lesson report. All of the homework activities are designed to be completed independently but if you would like to help with completion of homework, your tutor will be able to tell you how you can help.

FURTHER SUPPORT

Parents and guardians often ask a tutor what else they can do to support their child's learning or what resources they can buy to provide extra revision and practice. As a Pearson resource, *Tutors' Guild* has been designed to complement the popular *Revise* series.
Useful titles you may wish to purchase include:

- *Revise* AQA GCSE (9–1) Mathematics Foundation Revision Guide (ISBN: 9781447988069)
- *Revise* AQA GCSE (9–1) Mathematics Foundation Revision Workbook (ISBN: 9781447987864)
- *Revise* AQA GCSE (9–1) Mathematics Foundation Revision Cards (ISBN: 9781292182094)

Using pages 11–12 of this pack, your tutor will be able to tell you which pages of these *Revise* resources are appropriate for each lesson. If you purchase a set of Revision Cards, each card has a page reference in the top corner.

MATHS
— FOUNDATION —

INFORMATION FOR PARENTS AND GUARDIANS

WHAT'S IN THE TEST?

You may have heard a lot about the new GCSE (9–1) qualifications from your child's school, from other parents or in the media. Here is a breakdown of the AQA GCSE (9–1) Mathematics exams.

THE PAPERS

Your son or daughter will sit three GCSE (9–1) Mathematics papers. The exam is tiered, which means all candidates sit either Foundation or Higher papers. Each paper is worth 80 marks – one third of the total marks available. Candidates are given 1 hour and 30 minutes to complete each paper.

Paper 1 is non-calculator, while calculators are allowed in Papers 2 and 3. The GCSE (9–1) qualifications see an increase in the amount of calculator-allowed assessment, from 50% to 66.6%.

ASSESSMENT OBJECTIVES

There are three broad types of skills that are tested in the exams. Each one has its own *Assessment Objective*.

Assessment Objective 1 (AO1) tests how well a candidate can apply standard techniques. AO1 is the most accessible objective for the majority of candidates and could test a range of content areas, from ordering decimals, fractions and percentages to calculating with roots and powers. Around 50% of the marks available on Foundation papers and 40% of the marks available on Higher papers are awarded for AO1.

Assessment Objective 2 (AO2) assesses reasoning, interpretation and communication in Mathematics. It involves making inferences and conclusions, presenting evidence and proving answers. Trialling of the new exams found that many candidates find AO2 difficult to access. AO2 accounts for 25% of the Foundation marks and 30% of the Higher marks.

Assessment Objective 3 (AO3) assesses problem-solving skills. Based on recent trials of the new exams, this seems to be the area that causes candidates the most difficulty. However, mastering problem-solving could prove hugely beneficial, as it is worth 25% of the total marks for Foundation students and 30% for Higher students. The GCSE (9–1) Mathematics area of the AQA website has a range of free resources to help strengthen these skills.

RESULTS AND GRADES

GCSE results day is typically the third or fourth Thursday in August. It is the same day across the country, so you can find out the exact date online. On results day, the student will be given a slip of paper (or one per exam board, if the school hasn't collated them) with an overall grade for each GCSE. Grades for the GCSE qualifications are no longer given as letters (A*–U) but as numbers (9–1) instead. The diagram below shows roughly how the old-style grades translate to the new ones.

Previous grade	A*		A		B		C	D	E		F	G	U
New grade	9	8		7	6	5		4	3	2		1		U

As you can see, the new grade 9 is pitched higher than an A*. There is a wider spread of grades available for students whose target would previously have been a B/C. Because Mathematics is tiered, Foundation students will be able to access grades 1–5, whereas Higher students should be aiming to achieve grades 4–9.

SPECIFICATION GUIDANCE

The new AQA GCSE (9–1) Mathematics specification was introduced in 2015, with first assessment in 2017. If you have experience in tutoring or teaching the previous curriculum, much of the content and assessment will be familiar. If this is the case, please turn to pages 8–10 for guidance on what has changed.

If you are new to tutoring GCSE (9–1) Mathematics, this page will give you a brief introduction before you move on to pages 8–10. Further guidance on specific areas of the specification – including common misconceptions and barriers to learning – can be found in the lesson plans throughout this pack. The complete specification can be found on the AQA website.

KEY FACTS

Content domains

There are six areas of mathematics that will be assessed:

1. Number
2. Algebra
3. Ratio, proportion and rates of change
4. Geometry and measures
5. Probability*
6. Statistics*

The weighting of these content domains will be approximately as follows.

Content domain	Foundation	Higher
Number	25%	15%
Algebra	20%	30%
Ratio, proportion and rates of change	25%	20%
Geometry and measures	15%	20%
Statistics and probability*	15%	15%

*Statistics and probability are often considered together in assessment weightings.

The six domains are further broken down into smaller content areas. These are set out from page 10 onwards of the AQA specification which can be found on AQA's website. Foundation students will need to know all of the content identified in the specification in the basic foundation content and additional foundation content columns.

Each lesson plan in this pack highlights which areas of the specification are covered. In order to maximise your student's chances of success, the pack covers the most important specification areas and those students struggle with the most; it is intended to supplement and enhance classroom teaching and does not therefore cover the entire specification.

Exam papers

Both Foundation and Higher students will sit three exam papers:

- Paper 1; non-calculator; 80 marks; 1.5 hours
- Paper 2; calculator; 80 marks; 1.5 hours
- Paper 3; calculator; 80 marks; 1.5 hours

The six content domains will be assessed on all three of the papers. As the breakdown above shows, each exam paper is given an equal weighting: one third of the total available marks is available for each paper.

Foundation students will be able to access grades 1–5, whereas Higher students should be aiming to achieve grades 4–9. You will find further information about the format and structure of the exams and the new grading system in the corresponding *Assessment Pack* for this title (ISBN: 9781292195551).

MATHS
— FOUNDATION —

SPECIFICATION GUIDANCE

WHAT'S CHANGED?

Key changes

There are several key changes to the AQA GCSE (9–1) Mathematics course, brought about by new Ofqual requirements. The hope is that increasing the demand of GCSE Mathematics will better prepare students to apply their learning in everyday life, in work and in further studies. The new course will be more demanding in the following ways.

- **There is more subject content.**
  You'll have more topics to cover with your student, and the topics will be denser. This will change how you teach the course: will you recommend increased contact time, set more independent work or prioritise which content you cover?
- **Content is more demanding.**
  Both Foundation and Higher students will face more difficult topics than before.
- **Students will need to memorise more formulae.**
  Fewer formulae will be available to students in the exams. The formulae needed can be found on page 10.
- **There are more questions covering difficult skills.**
  Assessment objectives testing problem solving and reasoning carry more marks than previously for both Foundation and Higher students. These skills present difficulties for many students.
- **There are more marks to be gained for more difficult questions.**
  For both Foundation and Higher students, fewer marks will be available for lower-grade questions and more marks will be attributed to the questions at the top end of the grade scale.
- **Total exam time has increased.**
  There are now three exam papers to sit rather than the previous two, taking total exam time from 3.5 hours to 4.5 hours. This may present additional problems for students with low attention or concentration levels.

New content

For Foundation students, there is a large number of new topics that, in the previous GCSE specification, were Higher tier only. New topics include the following:
- compound interest and reverse percentages
- direct and indirect proportion
- standard form.

If you exclusively teach Foundation students, it may have been a while since you have come across this content and, as such, you may wish to spend some time refamiliarising yourself with it before tuition begins.

For Higher students, there are some new topics at the top end of the grade scale. These are intended to stretch students who go on to study A level. Such topics include the following:
- expanding the products or more than two binomials
- calculating or estimating gradients of graphs and areas under graphs, and interpreting results in real-life cases
- deducing turning points by completing the square.

To make way for this new content, some previously covered topics have been omitted from the new specification, including: 3D co-ordinates, imperial units of measure, and tessellations.

While this pack will help you to deliver the new content, you should make sure you are familiar and comfortable with the new topics and best practices for teaching them.

A full list of new content can be found on AQA's website.

SPECIFICATION GUIDANCE

REASONING AND PROBLEM SOLVING SKILLS

The new GCSE (9–1) qualification places increased emphasis on the more involved skills of reasoning and problem-solving than previously, so you need to make sure you are able to help your students develop such skills.

Reasoning, interpreting and communication mathematically

These skills are covered by the Assessment Objective AO2. Quality of Written Communication (QWC) is also assessed within AO2. There are no longer marks allocated to QWC, but it is taken into account when considering the effectiveness of a candidate's communication skills.

Questions that will help you to assess your student's reasoning ability will use commands such as *show that* and will involve setting out working in a clear, methodical way. It is worth reinforcing to your student that, in Maths, communicating well and explaining how you've arrived at a solution doesn't mean writing long paragraphs or even full sentences – it's about clear and methodical working, supported by short passages of text where necessary.

Linked to this is the requirement to be able to present arguments effectively or assess the validity of a given argument. Questions testing this skill come in various forms: students may, for example, need to explain why a statement is wrong or answer a question and then provide supporting evidence. Students will often struggle to know what an adequate response to such questions will be, so your support here will be valuable. You could look at acceptable responses in mark schemes or exemplar material with your student, or simply to work through questions orally so that you can guide their response.

Students may also be required to evaluate the presentation of information, such as a chart or table. Perhaps the most useful preparation here is to emphasise best practice in your student's own work and encourage the review of work with a critical eye.

To assess your student's ability to interpret information accurately, you can use questions that rely on the use and manipulation of geometric and graphical information. For example, a question that asks your student to calculate a value for the mean from a frequency table requires a solid understanding of the data that are presented (are they looking at cumulative or grouped frequency, for example?) and how that data should be interpreted to find the correct value.

Problem-solving

The proportion of marks awarded to problem-solving questions has been increased for the GCSE (9–1) qualifications. Previously, both Foundation and Higher exams awarded between 15% and 25% of marks for problem-solving, which is referred to as AO3 in the specification. For the GCSE (9–1) qualifications, Foundation and Higher papers carry approximately 25% and 30% AO3 marks respectively.

Trials have found that, in general, problem-solving is the area of Maths that students find the most difficult. For many, this will be because problem-solving questions often require a different approach from most other mathematical questions. The skills that make up AO3 include the following:

- **translating problems from mathematical or non-mathematical contexts into a mathematical process.**
  This essentially means taking a problem and working out what maths needs to be done to solve it.
- **making and utilising connections between different mathematical topics.**
  For example, recognising that algebra can help to solve probability problems and being able to put that into practice.
- **contextual interpretation of results.**
  For example, the logical application of units to an answer or calculation, or appropriate rounding of an answer to account for context (generally, a decimal answer to *What is the capacity of the stadium?* would be inappropriate).
- **evaluation of methods used, results obtained and the effect any assumptions may have had.**
  This could involve assessing the effectiveness or accuracy of two different methods, or commenting on the effect on a solution of not having access to complete information.

There are clear steps you can follow with your student to help them gain confidence in problem-solving: first, ensure they understand the problem; then, help them to formulate a plan; then, they can carefully carry out that plan; finally, they should check their answer logically and mathematically.

MATHS
— FOUNDATION —

AQA

SPECIFICATION GUIDANCE

REQUIRED FORMULAE

Foundation students need to know the following:	Higher students need to know the following:
Area rectangle = $l \times w$ parallelogram = $b \times h$ triangle = $\frac{1}{2}b \times h$ trapezium = $\frac{1}{2}(a+b)h$	**Area** rectangle = $l \times w$ parallelogram = $b \times h$ triangle = $\frac{1}{2}b \times h$ trapezium = $\frac{1}{2}(a+b)h$
Volumes cuboid = $l \times w \times h$ cylinder = $\pi r^2 h$ prism = area of cross section × length	**Volumes** cuboid = $l \times w \times h$ cylinder = $\pi r^2 h$ prism = area of cross section × length pyramid = $\frac{1}{3}$ × area of base × h
Circles circumference = π × diameter $C = \pi d$ circumference = 2 × π × radius $C = 2\pi r$ area of circle = π × radius squared $A = \pi r^2$	**Circles** circumference = π × diameter $C = \pi d$ circumference = 2 × π × radius $C = 2\pi r$ area of circle = π × radius squared $A = \pi r^2$
Compound measures speed = $\dfrac{\text{distance}}{\text{time}}$ density = $\dfrac{\text{mass}}{\text{volume}}$ The formula for pressure will be provided if relevant.	**Compound measures** speed = $\dfrac{\text{distance}}{\text{time}}$ density = $\dfrac{\text{mass}}{\text{volume}}$ The formula for pressure will be provided if relevant.
Pythagoras Pythagoras' Theorem: for a right-angled triangle, $a^2 + b^2 = c^2$ trigonometric ratios: $\sin x = \dfrac{\text{opp}}{\text{hyp}}$ $\cos x = \dfrac{\text{adj}}{\text{hyp}}$ $\tan x = \dfrac{\text{opp}}{\text{adj}}$	**Pythagoras** Pythagoras' Theorem: for a right-angled triangle, $a^2 + b^2 = c^2$ trigonometric ratios: $\sin x = \dfrac{\text{opp}}{\text{hyp}}$ $\cos x = \dfrac{\text{adj}}{\text{hyp}}$ $\tan x = \dfrac{\text{opp}}{\text{adj}}$
	Trigonometric formulae sine rule $\dfrac{a}{\sin A} = \dfrac{b}{\sin B} = \dfrac{c}{\sin C}$ cosine rule $a^2 = b^2 + c^2 - 2bc \cos A$ area of triangle = $\frac{1}{2} ab \sin C$
	Quadratic equations The solutions of $ax^2 + bx + c = 0$, where $a \neq 0$, Are given by $x = -b \pm \sqrt{\dfrac{b^2 - 4ac}{2a}}$

These formulae are listed by AQA as need-to-know.

REVISE MAPPING GUIDE

Pearson's *Revise* series provides simple, clear support to students preparing for their GCSE (9–1) exams. Parents may ask you if you know of any independent study resources that they can work through with their child, or you may wish to provide such resources yourself.

We have provided below a mapping guide for each lesson in this pack to a corresponding page in the *Revise* series, to make such recommendations easier for you.

For students studying AQA GCSE (9-1) Mathematics Foundation, we recommend the following titles:
- *Revise* AQA GCSE (9–1) Mathematics Foundation Revision Guide (ISBN: 9781447988069)
- *Revise* AQA GCSE (9–1) Mathematics Foundation Revision Workbook (ISBN: 9781447987864)
- *Revise* AQA GCSE (9–1) Mathematics Foundation Revision Cards (ISBN: 9781292182094)

The Revision Guides and Revision Workbooks correspond page-for-page, so the page references are the same for both, and each Revision Card has the page reference in the top corner.

REVISE AQA GCSE (9–1) MATHEMATICS FOUNDATION REVISION GUIDE AND REVISION WORKBOOK

LESSON		WHAT'S IN THE BOOK?	PAGES
1	Diagnostic lesson	Find out what the student knows; Find out their preferences and attitudes	
2	Calculations	Squares, cubes and roots, Indices	8, 9
3	Standard form	Operations on decimals; Standard form 2	17, 18
4	Fractions, decimals and percentages	Decimals and place value; Operations on decimals; Fractions; Operations on fractions; Mixed numbers; Percentages; Fractions, decimals and percentages; Percentage change 1; Percentage change 2; Reverse percentages; Growth and decay	6, 7, 13, 14, 15, 56, 57, 58, 59, 63, 64
5	Rounding and error intervals	Rounding numbers; Estimation; Inequalities	3, 10, 33
6	Estimating	Operations on decimals; Estimation	7, 10
7	Straight line graphs	Gradients of lines; Straight-line graphs 1; Straight-line graphs 2	38, 39, 40
8	Quadratic, cubic and reciprocal graphs	Quadratic graphs; Using quadratic graphs; Cubic and reciprocal graphs	45, 46, 49
9	Interpreting graphs	Real-life graphs; Distance–time graphs; Rates of change	41, 42, 43
10	Solving linear equations	Collecting like terms Simplifying expressions; Substitution; Linear equations 1; Linear equations 2	22, 23, 25 30, 31
11	Quadratic equations	Factorising; Factorising quadratics; Quadratic equations	29, 47, 48
12	Simultaneous equations	Simultaneous equations	50
13	Inequalities	Inequalities; Solving inequalities	33, 34
14	Sequences	Sequences 1; Sequences 2	35, 36
15	Ratio	Ratio 1; Ratio 2	60, 61
16	Direct and inverse proportion	Proportion; Proportion and graphs	68, 69
17	Proportion graphs	Distance–time graphs; Rates of change; Growth and decay; Proportion and graphs	42, 43, 64, 69
18	Angles	Angles 1; Angles 2; Solving angle problems; Angles in polygons; Measuring and drawing angles	74, 75, 76, 77, 96
19	Ruler and compass construction	Constructions 1; Constructions 2; Loci	100, 101, 102

REVISE MAPPING GUIDE

REVISE AQA GCSE (9–1) Mathematics Foundation Revision Guide
AND Revision Workbook

Lesson		What's in the book?	Pages
20	Circles	Circles; Area of a circle; Sectors of circles	104, 105, 106
21	Pythagoras' theorem	Pythagoras' theorem; Line segments	91, 92
22	Using trigonometry	Trigonometry 1; Trigonometry 2; Solving trigonometry problems	93, 94, 95
23	Trigonometric values	Solving trigonometry problems	95
24	Properties of shapes	Symmetry; Quadrilaterals; Angles 1	72, 73, 74
25	Congruent and similar shapes	Similarity and congruence; Similar shapes; Congruent triangles	110, 111, 112
26	Transformations	Coordinates; Translations; Reflections; Rotations; Enlargements	37, 87, 88, 89, 90
27	Bearings and scale	Scale drawings and maps; Bearings	99, 103
28	3D shapes	3D shapes; Plans and elevations	83, 98
29	Area and volume	Perimeter and area; Area formulae; Units of area and volume; Volumes of cuboids; Area of a circle; Volumes of 3D shapes; Surface area	80, 81, 86, 105, 108, 109
30	Introduction to vectors	Vectors	113
31	Calculating with vectors	Vectors	113
32	Basic probability and Venn diagrams	Probability 1; Venn diagrams	127, 131
33	Combined probability	Probability 2; Independent events	128, 132
34	Planning an investigation and data collection	Sampling	125
35	Constructing graphs, charts and diagrams	Pictograms; Bar charts; Pie charts; Line graphs;	117, 118, 119, 124
36	Measures of spread and location	Averages and range; Comparing data	121, 126
37	Using measures of location and spread	Averages from tables 1; Averages from tables 2	122, 123
38	Scatter graphs	Scatter graphs	120

NEEDS ANALYSIS

FOR PARENTS AND GUARDIANS

We have a tutor because ...
(Briefly explain why you have employed a tutor.)

Where we are currently ...
(Briefly explain the student's current progress. Do you have access to reports and predicted grades?)

FOR STUDENTS

Use this space to tell your tutor about yourself.

I am ...
Tell your tutor what type of person you think you are. Are you quiet or outgoing? Are you confident about your abilities?

I like...
Explain to your tutor how you like to work. Do you like to work independently or with more guidance? Do you like to write your answers down or talk through them first? Do you like to be creative?

How I feel about maths ...
Do you like maths? Try to explain why or why not. What are your favourite and least favourite parts?

NEEDS ANALYSIS

OUR GOALS

Work together to set small, achievable goals for the year ahead. Make them as positive as you can and don't limit your goals to areas of maths – think about personal development too. Together, look back at this list often to see how you are progressing.

TICK OFF EACH GOAL WHEN YOU'VE ACHIEVED IT

In four weeks' time, I will …

☐ ....

☐ ....

☐ ....

☐ ....

☐ ....

☐ ....

In three months' time, I will …

☐ ....

☐ ....

☐ ....

☐ ....

☐ ....

☐ ....

By the time I sit my exam, I will …

☐ ....

☐ ....

☐ ....

☐ ....

☐ ....

☐ ....

PROGRESS REPORT

Fill in the boxes below with help from your tutor.

My strengths are ...
Which areas of maths do you think you've done well in recently? List at least three.

My favourite maths topic is ...
Which maths topic is your favourite? It doesn't have to be the one you're best at!

because ...

The areas of maths I need to work on are ...
In which areas of maths do you think you need more practice?

To improve these areas, we are going to ...
This space is for your tutor to explain how he/she is going to help you become confident in these areas.

END-OF-LESSON REPORT

We have looked at last week's homework and my tutor thinks …
This space is for your tutor to give feedback on last week's homework.

Today, we worked on …
This space is for you to list all of the topics and skills that you and your tutor have worked on today.

I feel …
This space is for you to explain how you feel about today's lesson. Did you enjoy it? Do you feel confident?

My tutor thinks …
This space is for your tutor to explain how the lesson went.

At home this week, we can…
This space is for your tutor to explain what your homework is and give you other ideas for extra revision and practice.

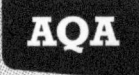 

1 DIAGNOSTIC LESSON

LEARNING OBJECTIVES

- Know the properties of 2D and 3D shapes
- Add and subtract decimal numbers
- Multiply a decimal by a whole number
- Multiply 3-digit by 2-digit whole numbers
- Use formal methods of multiplication and understand place value

SPECIFICATION LINKS

- G1, G4, G6, N2

STARTER ACTIVITY

- **Odd one out; 5 minutes; page 18**
  For each of the sets of shapes shown, ask the student to decide which is the odd one out and to explain why. Encourage the use of mathematical language and explain that there is more than one possible odd one out.

MAIN ACTIVITIES

- **Win or lose?; 20 minutes; page 19**
  Full instructions are given on the activity sheet. You may wish to play each game several times before discussing strategies. Encourage the student to use appropriate mathematical language.
- **More than one method; 20 minutes; page 20**
  Before starting this activity, discuss how each of the multiplication methods works. Encourage the student to communicate their reasoning to you. For example, for question 2 ask *'how do you know what 127 × 4 is equal to?'*

PLENARY ACTIVITY

- **Three things I've learnt; 5 minutes**
  Ask the student to describe three things they have learnt from this session. Encourage them to reflect on their learning and include statements like *'it's okay to ask questions'*.

HOMEWORK ACTIVITY

- **Self-assessment; 20 minutes; page 21**
  Explain to the student that the success of tutoring relies on what both parties put into it. With this in mind, explain how to complete the activity and ask the student to ensure they spend time thinking about their answers, as this will give you an idea of how best to support them.

SUPPORT IDEA

- **More than one method** Work through both methods of multiplication, explaining how they work before attempting the questions. You could simplify the multiplications to 127 × 4, ensuring you also simplify the linked questions.

EXTENSION IDEA

- **More than one method** Challenge the student to write down as many related facts as they can using the traditional long multiplication calculation.

PROGRESS AND OBSERVATIONS

— FOUNDATION —

STARTER ACTIVITY: ODD ONE OUT

TIMING: 5 MINS

LEARNING OBJECTIVES

- Know the properties of 2D and 3D shapes

EQUIPMENT

1. Which shape is the odd one out? Explain your answer.

a)

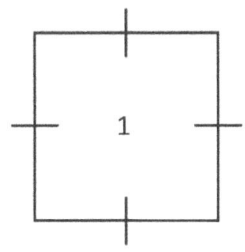 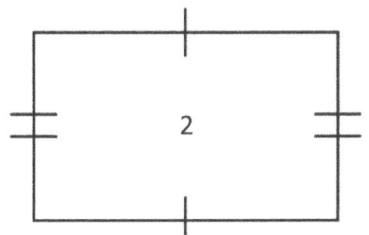 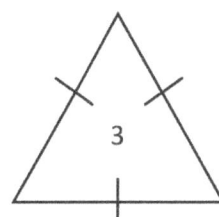

b)

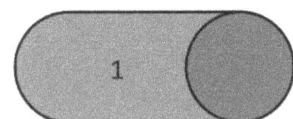 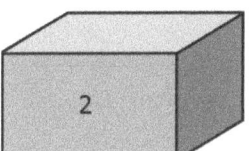 

c)

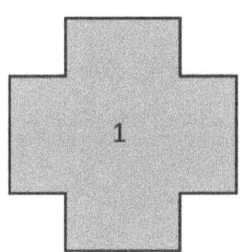 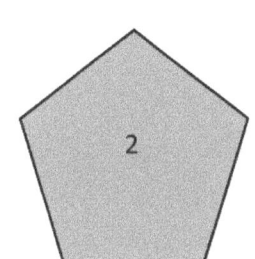 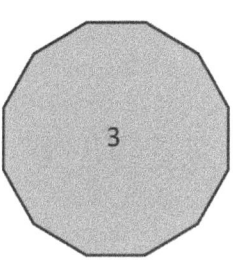

MAIN ACTIVITY: WIN OR LOSE?	TIMING: 20 MINS

LEARNING OBJECTIVES
- Add and subtract decimal numbers
- Multiply a decimal by a whole number

EQUIPMENT
- dice

1. **Take it in turns to roll a dice. You and your tutor should both write every number rolled into one of the squares in the additions below. The winner is the person with the highest answer when all six numbers have been rolled and written in. On each answer line, describe the strategy you used to try and win.**

2. **Repeat the game for these subtractions.**

3. **Repeat the game for these multiplications.**

MAIN ACTIVITY: MORE THAN ONE METHOD   TIMING: 20 MINS

LEARNING OBJECTIVES
- Multiply 3-digit by 2-digit whole numbers
- Use formal methods of multiplication and understand place value

EQUIPMENT
- calculator

Ethan and Amina have both worked out 24 × 127 using different methods.

Ethan

```
        1   2   7
    ×       2   4
    ─────────────
        5   0   8
        ¹   ²
    2   5   4   0
        ¹
    ─────────────
    3   0   4   8
    ¹
```

Amina

×	100	20	7
20	2000	400	140
4	400	80	28

```
    2   0   0   0
    4   0   0
    1   4   0
    4   0   0
        8   0
+       2   8
─────────────
3   0   4   8
¹   ¹
```

1. **Which method do you prefer? Explain why.**

----

2. **Use Ethan's working out to write down the answers to these calculations.**

   a) 127 × 4 = ...

   b) 127 × 20 = ...

   c) 3048 − 508 = ...

3. **Use Amina's working out to write down the answers to these calculations.**

   a) 20 × 20 = ...

   b) 24 × 7 = ...

   c) 3048 ÷ 24 = ...

4. **Given that 127 × 24 = 3048, write down the answers to these calculations. Explain to your tutor how you found each answer.**

   a) 12.7 × 24 = ...

   b) 0.24 × 127 = ...

   c) 3048 ÷ 1270 = ...

   d) 0.127 × 24 = ...

   e) 3048 ÷ 2.4 = ...

   f) 30.48 ÷ 127 = ...

HOMEWORK ACTIVITY: SELF-ASSESSMENT   TIMING: 20 MINS

LEARNING OBJECTIVES

* Self-assessment of strengths and weaknesses

EQUIPMENT

1. One of the ways you can develop your mathematical skills is to recognise your own weaknesses. Think carefully about the answers to each of these questions and answer them truthfully. Decide if you agree or disagree with the statements – you have four options!

	Strongly agree	Agree	Disagree	Strongly disagree
I find all parts of maths difficult.				
I often make mistakes in my calculations.				
I do not read the questions carefully.				
I get bored in maths and stop listening.				
I am trying as hard as I can in maths.				
I don't want to ask questions in case I look foolish.				
I understand it in the lesson, but when I have to do it on my own I find it tricky.				
I find problem-solving difficult.				
In maths tests, I don't understand what the questions are asking.				
I don't remember the methods I have been taught.				
There are some topics in maths that I have never understood.				
I don't understand why I need to do some topics in maths.				

If you agreed with either of the last two statements, write down the topics you were thinking of.

1 ANSWERS

STARTER ACTIVITY: ODD ONE OUT

1. Accept any correct answers. For example:
a) 2 (irregular) or 3 (no right angles / not a quadrilateral)
b) 2 (no curved faces) or 3 (no edges)
c) 1 (irregular) or 2 (not a dodecagon)

MAIN ACTIVITY: WIN OR LOSE?

1–3. Check student's answers.

MAIN ACTIVITY: MORE THAN ONE METHOD

1. Student's own answer
2. a) 508   b) 2540   c) 2540
3. a) 400   b) 168   c) 127
4. a) 304.8   b) 30.48   c) 2.4   d) 3.048   e) 1270   f) 0.24

HOMEWORK ACTIVITY: SELF-ASSESSMENT

1. Student's own answers. Discuss the answers with the student.

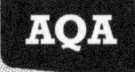 
2 NUMBER: CALCULATIONS

LEARNING OBJECTIVES

- Recognise powers and roots of whole numbers
- Use positive integer powers and roots
- Calculate with roots and integer indices
- Calculate exactly with fractions and multiples of π

SPECIFICATION LINKS

- N2, N3, N4, N6, N7, N8

STARTER ACTIVITY

- **Heads and tails; 5 minutes; page 24**
  Ask students to match the 'head' and the 'tail' in the image.

MAIN ACTIVITIES

- **Calculating exactly; 20 minutes; page 25**
  Explain that we sometimes need to leave solutions in exact terms. This might include the use of fractions and π. Remind the student how to convert between a mixed number and an improper fraction, and establish how to carry out the four operations with fractions (include using cross cancelling first to simplify calculations for multiplication and division). For question 3, stress that π is an exact value. It can be treated in the same way as an algebraic term.

- **Laws of indices; 20 minutes; page 26**
  Remind the student of the meaning of index notation and particularly of the two laws: $a^1 = a$ and $a^0 = 1$. To avoid the common misconception that $a^0 = 0$ you may wish to model some examples of the form $\dfrac{a \times \cancel{a} \times \cancel{a} \times \cancel{a} \times \cancel{a}}{a \times \cancel{a} \times \cancel{a} \times \cancel{a} \times \cancel{a}} = \dfrac{a}{a}$ and use this to prove that any number divided by itself equals 1. Revise the idea of a reciprocal, modelling finding the reciprocal of a decimal as well as of whole numbers. Discuss how a number can be raised to a negative power, and model some examples of this.

PLENARY ACTIVITY

- **The four operations with fractions; 5 minutes**
  Ask the student to draw flow charts to show how to add, subtract, multiply and divide fractions.

HOMEWORK ACTIVITY

- **Quick quiz; 30 minutes; page 27**
  Full instructions are given on the activity sheet.

SUPPORT IDEAS

- **Calculating exactly** For question 1, start with fractions that are not mixed numbers, working through some examples before asking them to try the activity.
- **Laws of indices** Expand the expressions (e.g. $3^4 \times 3^8 = 3 \times 3 \times 3 \times 3 \times 3 \times 3 \times 3 \times 3 \times 3 \times 3 \times 3 \times 3$).

EXTENSION IDEA

- **Calculating exactly** Challenge the student to find the largest possible answer for question 1 part a), given that all six numbers in the calculation are different ($5\frac{2}{3} + 6\frac{1}{4}$ or $6\frac{2}{3} + 5\frac{1}{4} = 11\frac{11}{12}$).

PROGRESS AND OBSERVATIONS

STARTER ACTIVITY: HEADS AND TAILS

TIMING: 5 MINS

LEARNING OBJECTIVES

- Recognise powers and roots of whole numbers

EQUIPMENT

1. Ali is making coins. The values on the heads and tails sides of each coin must be equal. Draw lines to match up the heads (dark grey) and tails (light grey) to show the pairs of values he should write on each coin.

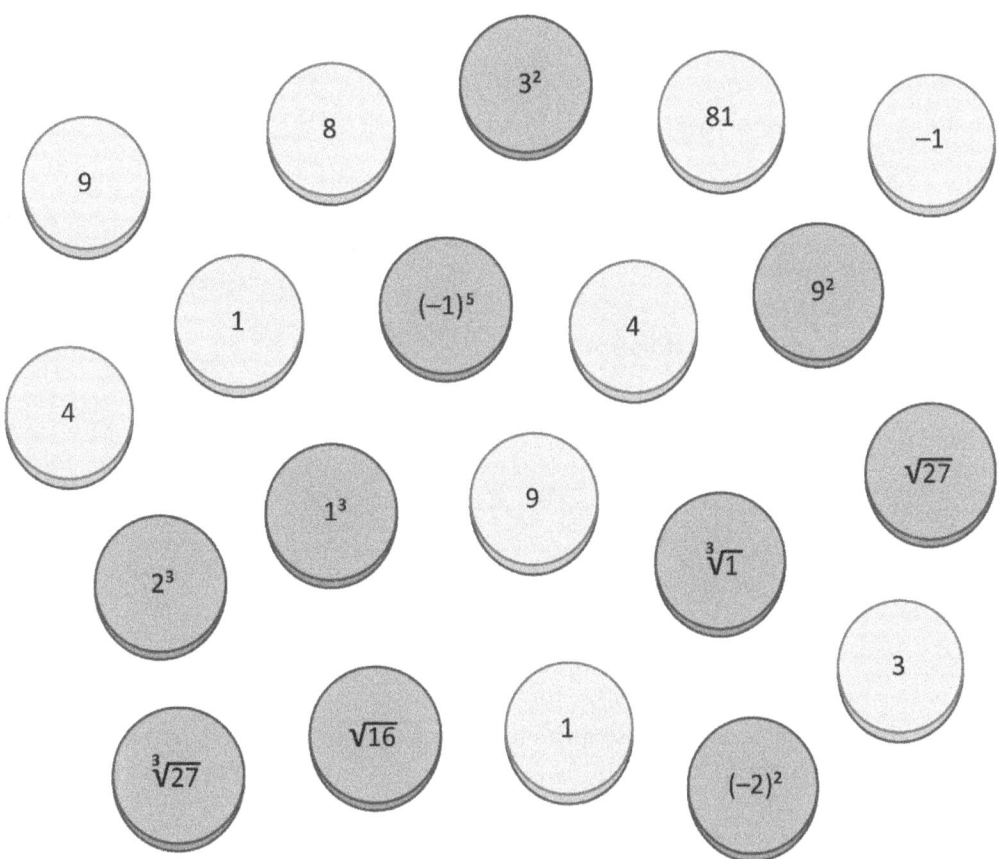

MAIN ACTIVITY: CALCULATING EXACTLY   TIMING: 20 MINS

LEARNING OBJECTIVES
- Calculate exactly with fractions and multiples of π

EQUIPMENT
- three dice

1. **Roll three dice to make a mixed number. For example, if you rolled a two, a three and a five, you could make the mixed number** $3\frac{2}{5}$. **Your numerator must not be larger than your denominator.**

   Do this twice to create two mixed numbers. Without using a calculator:

   a) add together the two values   ....

   b) subtract the smaller value from the larger   ....

   c) multiply the two values together   ....

   d) divide the larger value by the smaller value.   ....

2. **Work out the area and perimeter of the following shapes.**

   a) a square with sides of length $\frac{1}{4}$ cm   ....

   b) a rectangle of width $\frac{3}{4}$ cm and length $1\frac{1}{3}$ cm   ....

3. **Write down an expression for:**

   a) 2 more than π   ....

   b) 5 less than π   ....

   c) four times π   ....

   d) one third of π.   ....

4. **The circumference of a circle is calculated using the formula 2πr. A circle has a circumference of kπ. Its diameter is 12 cm. What is the value of k?**

   ....

5. **The formula for the area of a circle is πr^2. Work out the exact area of a circle with:**

   a) a radius of 3 cm   ....

   b) a radius of $\frac{1}{2}$ an inch   ....

   c) a diameter of 14 cm.   ....

MAIN ACTIVITY: LAWS OF INDICES   **TIMING: 20 MINS**

LEARNING OBJECTIVES

- Use +ve integer powers and roots
- Calculate with integer indices

EQUIPMENT

Complete these questions without using a calculator.

1. **Write these expressions as a power of 3.**

   a) $3 \times 3 \times 3 \times 3$ ....

   b) $3 \times 3 \times 3 \times 3 \times 3 \times 3$ ....

   c) 1 ....

   d) 3 ....

2. **Work out the value of these expressions.**

   a) 2^4 ....

   b) 1^8 ....

   c) $(-2)^2$ ....

   d) $(-3)^3$ ....

   e) $\left(\dfrac{2}{3}\right)^2$ ....

3. **Give the reciprocals of these numbers.**

   a) 7   ....

   b) 0.25   ....

   c) $\dfrac{1}{3}$ ....

   d) $\dfrac{3}{4}$ ....

4. **Complete this table.**

2^3	2^2	2	2^0	2^{-1}	2^{-2}	2^{-3}	2^{-4}
$= 2 \times 2 \times 2$					$= \dfrac{1}{2 \times 2}$		
$= 8$	$= 4$			$= \dfrac{1}{2}$			

5. **Use the pattern you discovered in question 4 to work out the value of these expressions.**

   a) 3^{-2} ....

   b) 5^{-3} ....

   c) 6^{-1} ....

   d) 1^{-7} ....

6. **Pippa is practising with index rules. To calculate $3^2 \times 3^3$, she writes out: $3^2 \times 3^3 = (3 \times 3) \times (3 \times 3 \times 3) = 3^a$.**

   a) What is the value of a?   ....

   b) Pippa then writes out: $\dfrac{3^5}{3^2} = \dfrac{3 \times 3 \times 3 \times 3 \times 3}{3 \times 3} = 3^b$. What is the value of b?   ....

   c) Pippa then writes: $(3^2)^4 = (3 \times 3) \times (3 \times 3) \times (3 \times 3) \times (3 \times 3) = 3^c$. What is the value of c?   ....

HOMEWORK ACTIVITY: QUICK QUIZ   TIMING: 30 MINS

LEARNING OBJECTIVES

- Use +ve integer powers and roots
- Calculate with roots and integer indices
- Calculate with fractions and multiples of π

EQUIPMENT

1. **Explain how to carry out each of these operations in just one line.**

   a) adding fractions

   ....

   b) subtracting fractions

   ....

   c) multiplying fractions

   ....

   d) dividing fractions

   ....

2. **Work out the answer to each of these indices questions.**

   a) What is the value of $3^2 \times 3^5$?

   ....

   b) What is the value of $5^5 \div 5^2$?

   ....

   c) Sort these values, smallest first: 4^3, 4^0, 4^{-2}, $(4^2)^3$, 4^{-1}

   ....

   d) Sort these values, smallest first: $\sqrt{1}$, $\sqrt[3]{8}$, 3^{-1}, 4^0

   ....

   e) Sort these values, smallest first: $2^4 \times 2^2$, $\dfrac{2^{11}}{2^7}$, $(2^3)^3$

   ....

2 ANSWERS

STARTER ACTIVITY: HEADS AND TAILS

1. $3^2 = 9$ $9^2 = 81$ $1^3 = 1$ $\sqrt{81} = 9$ $2^3 = 8$

   $(-2)^2 = 4$ $(-1)^5 = -1$ $\sqrt[3]{27} = 3$ $\sqrt{16} = 4$ $\sqrt[3]{1} = 1$

MAIN ACTIVITY: CALCULATING EXACTLY

1. Check student's answers.

2. a) perimeter = 1 cm, area = $\frac{1}{16}$ cm^2

   b) perimeter = $4\frac{1}{6}$ cm, area = 1 cm^2

3. a) $\pi + 2$ b) $\pi - 5$ c) 4π d) $\frac{\pi}{3}$

4. 12

5. a) 9π cm^2 b) $\frac{\pi}{4}$ inches^2 c) 49π cm^2

MAIN ACTIVITY: LAWS OF INDICES

1. a) 3^4 b) 3^6 c) 3^0 d) 3^1

2. a) 16   b) 1   c) 4   d) -27 e) $\frac{4}{9}$

3. a) $\frac{1}{7}$ b) 4   c) 3   d) $\frac{4}{3}$

4.

2^3	2^2	2	2^0	2^{-1}	2^{-2}	2^{-3}	2^{-4}
$= 2 \times 2 \times 2$	$= 2 \times 2$	$= 2$	$= 1$	$= \frac{1}{2}$	$= \frac{1}{2 \times 2}$	$= \frac{1}{2 \times 2 \times 2}$	$= \frac{1}{2 \times 2 \times 2 \times 2}$
$= 8$	$= 4$	$= 2$	$= 1$	$= \frac{1}{2}$	$= \frac{1}{4}$	$= \frac{1}{8}$	$= \frac{1}{16}$

5. a) $\frac{1}{9}$ b) $\frac{1}{125}$ c) $\frac{1}{6}$ d) 1

6. a) 5   b) 3   c) 8

HOMEWORK ACTIVITY: QUICK QUIZ

1. Check student's answers.

2. a) 3^7 b) 5^3 c) $4^{-2}, 4^{-1}, 4^0, 4^3, (4^2)^3$ d) $3^{-1}, [4^0, \sqrt{1}$ OR $\sqrt{1}, 4^0], \sqrt[3]{8}$ e) $\frac{2^{11}}{2^7}, 2^4 \times 2^2, (2^3)^3$

GLOSSARY

Index (plural indices)/power
The number that indicates how many times to multiply the base number by itself

3 NUMBER: STANDARD FORM

LEARNING OBJECTIVES

- Order positive numbers
- Calculate with and interpret standard form $A \times 10^n$, where $1 \le A < 10$ and n is an integer

SPECIFICATION LINKS

- N1, N9

STARTER ACTIVITY

- **Ordering numbers; 5 minutes; page 30**
  Ask the student to put the numbers in the cloud in ascending order. If necessary, encourage them to write the value of each number (e.g. $10^5 = 10\,000$) and to convert the fractions to decimals.

MAIN ACTIVITIES

- **Numbers in standard form; 20 minutes; page 31**
  Explain how to write very large and very small numbers in standard form, and how to convert these numbers back to ordinary numbers. Show the student how to enter a number in standard form into a calculator and how it will be represented on a calculator display.
- **Calculating with standard form; 20 minutes; page 32**
  Show the student how to carry out calculations involving standard form. Show how multiplication and division questions can be simplified by combining the powers of 10, but emphasise that for addition and subtraction, numbers in standard form must be written as ordinary numbers first (unless the power of 10 is the same).

PLENARY ACTIVITY

- **How to write in standard form; 5 minutes**
  Ask the student to explain, in fewer than 20 words, how to write a number in standard form.

HOMEWORK ACTIVITY

- **Exam-style questions; 45 minutes; page 33**
  Full information is provided on the activity sheet.

SUPPORT IDEAS

- **Numbers in standard form** Support the student by writing out fully the power of 10
  e.g. $3.2 \times 10^5 = 3.2 \times 10 \times 10 \times 10 \times 10 \times 10$
- **Calculating with standard form** Support the student by modelling how to add/subtract and multiply and divide expressions written in standard form. When multiplying and dividing numbers written in standard form, group together the powers of 10 and write them out fully, showing how to cancel. Remind the student that they must ensure their answer is written properly in standard form and that the numeric term is between 1 and 10.

EXTENSION IDEAS

- **Numbers in standard form** Extend by giving the student the following extension task:
  Fatima has completed her homework putting numbers into standard form, but she has managed to get every answer wrong! Explain what she has done wrong for each question.
  a) $45\,000 = 45 \times 10^3$ b) $0.00035 = 3.5 \times 10^5$ c) $0.109 = 1.9 \times 10^{-2}$ d) $1125 = 1.125 \times 10^4$
- **Calculating with standard form** Extend by asking the student to devise a 'rule' for simplifying a multiplication or division when using numbers written in standard form. Link this to the laws of indices.

PROGRESS AND OBSERVATIONS

MATHS
— FOUNDATION —

STARTER ACTIVITY: ORDERING NUMBERS

TIMING: 5 MINS

LEARNING OBJECTIVES

• Order positive numbers

EQUIPMENT

1. Write these numbers onto the number line in ascending order.

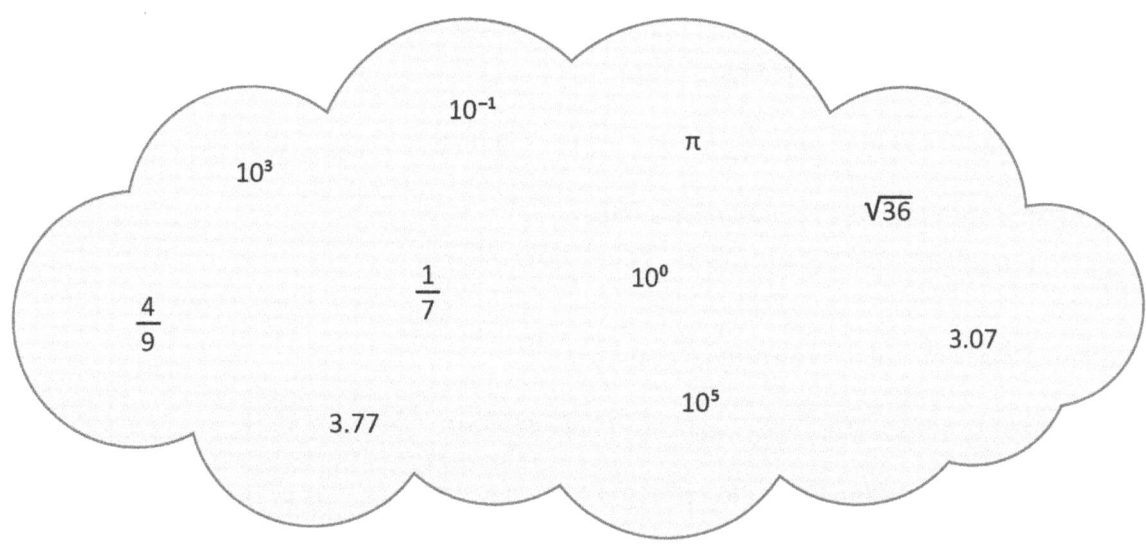

10^{-1}

π

10^3

$\sqrt{36}$

$\dfrac{1}{7}$

10^0

$\dfrac{4}{9}$

3.07

10^5

3.77

smallest   largest

MATHS
— FOUNDATION —

MAIN ACTIVITY: NUMBERS IN STANDARD FORM   **TIMING: 20 MINUTES**

LEARNING OBJECTIVES
- Write numbers in and interpret standard form

EQUIPMENT
- calculator

1. Write each of these large numbers in standard form.

   a) 34 000 000 000 .... b) 235 000 ....

2. Write each of these small numbers in standard form.

   a) 0.000000002 .... b) 0.00111 ....

3. Write each of these large numbers as an ordinary number.

   a) 3.4×10^5 .... b) 8.01×10^4 ....

4. Write each of these small numbers as an ordinary number.

   a) 4×10^{-2} .... b) 5.6×10^{-5} ....

5. Type the following calculations into your calculator and write down the answers as shown on the display. Convert them to ordinary numbers.

   Example:
   $$32\,000 \times 1\,900\,000 = 6.08 \times 10^{10}$$
   $$= 60\,800\,000\,000$$

   a) $450\,000 \times 923\,000\,000$ ....

   b) $1.7 \div 10\,000\,000 - 0.000000003$ ....

   c) 0.2^{17} ....

   d) $\sqrt{0.0000000009}$ ....

6. Put these numbers in order of size from smallest to largest without converting them to normal numbers first.

   3.8×10^{-2} 3.08×10^{-1} 3.8×10^4 3.81×10^8 3.88×10^{-5}

   ....

MATHS
— FOUNDATION —

AQA

MAIN ACTIVITY: CALCULATING WITH STANDARD FORM   TIMING: 20 MINS

LEARNING OBJECTIVES

- Calculate with standard form

EQUIPMENT

- calculator

1. The table below shows the average distance from the Sun to the planets in our solar system.
   Write the planets in order of distance from the Sun.

Planet	Saturn	Venus	Jupiter	Neptune	Mars	Uranus	Mercury	Earth
Average distance from the Sun (km)	1.4×10^9	1.08×10^8	7.78×10^8	4.49×10^9	2.30×10^8	2.97×10^9	5.8×10^7	1.50×10^8

closest   ....

....   furthest

2. Add together the distance from:

   a) the Sun to Mars and the Sun to Earth   ....

   b) the Sun to Venus and the Sun to Mercury. ....

3. Imagine that all the planets are in a straight line from the Sun. Work out the distance from:

   a) Uranus to Neptune   ....

   b) Mercury to Saturn.   ....

4. The speed of light is 1.08×10^9 km per hour.

   a) How long does it take (in hours) for one of the Sun's rays to reach Earth? ....

   b) Convert this time into minutes.   ....

5. The diameter of a bacterium is 1×10^{-4} cm. The diameter of the bacterium increases by 1×10^{-4} cm every minute. What is the diameter after the following amounts of time? Give your answer in standard form. Do not use a calculator.

   a) 1 minute   ....

   b) 10 minutes   ....

   c) 1 hour   ....

HOMEWORK ACTIVITY: EXAM-STYLE QUESTIONS

TIMING: 45 MINS

LEARNING OBJECTIVES

- Calculate with and interpret standard form $A \times 10^n$ where $1 \le A \le 10$ and n is an integer

EQUIPMENT

- calculator

Complete these questions without a calculator.

1. Write the number 450 000 in standard form. .... **(1 mark)**

2. Write the number 0.000034 in standard form. .... **(1 mark)**

3. Work out the value of $(8 \times 10^3) + (2 \times 10^4)$. Write your answer in standard form.

....
(2 marks)

4. Calculate the value of $(3 \times 10^2) \times (2 \times 10^3)$. Write your answer in standard form.

....
(2 marks)

5. Work out $(8 \times 10^{-2}) - (4 \times 10^{-2})$. Write your answer in standard form.

....
(2 marks)

6. Work out $(6 \times 10^5) \div (2 \times 10^3)$. Write your answer in standard form.

....
(2 marks)

You may use a calculator for the questions below, but you must show your working.

7. A piece of paper is 7×10^{-2} mm thick.

    a) How thick is a stack of 200 pieces of paper?

....
(2 marks)

    b) A photocopier tray is 20 cm deep. Paper comes in packs of 500 sheets. How many packs must you buy to ensure you can completely fill the photocopier tray?

....
(2 marks)

8. It is assumed the Earth is 4.6×10^9 years old. It is thought that the first men lived 2.09×10^5 years ago. If these facts are both true, how long did the Earth exist before man? Give your answer as an ordinary number, rounded to 5 significant figures.

....
(2 marks)

3 ANSWERS

STARTER ACTIVITY: ORDERING NUMBERS

1. 10^{-1} $\frac{1}{7}$ $\frac{4}{9}$ 10^0 3.07 π 3.77 $\sqrt{36}$ 10^3 10^5

MAIN ACTIVITY: NUMBERS IN STANDARD FORM

1. a) 3.4×10^{10} b) 2.35×10^5
2. a) 2×10^{-9} b) 1.11×10^{-3}
3. a) 340 000   b) 80 100
4. a) 0.04   b) 0.000 056
5. a) 4.1535×10^{14} = 415 350 000 000 000   b) 1.67×10^{-7} = 0.000000167
    c) 2×10^{16} = 20 000 000 000 000 000   d) 3×10^{-5} = 0.00003
6. 3.88×10^{-5} 3.8×10^{-2} 3.08×10^{-1} 3.8×10^4 3.81×10^8

EXTENSION TASK

a) 45 000 = 45×10^3: the 45 is not between 1 and 10
b) 0.00035 = 3.5×10^4: the power of 10 should be negative since the number is small
c) 0.109 = 1.9×10^{-2}: she has missed the zero between 1 and 9 – since this is a placeholder, it is essential
d) 1125 = 1.125×10^4: the power of 10 is one too large

MAIN ACTIVITY: CALCULATING WITH STANDARD FORM

1. Mercury, Venus, Earth, Mars, Jupiter, Saturn, Uranus, Neptune
2. a) 3.8×10^8 b) 1.66×10^8
3. a) 1.52×10^9 b) 1.342×10^9
4. a) $1.5 \times 10^8 \div 1.08 \times 10^9$ = 0.14 hours (2 d.p.)   b) $8\frac{1}{3}$ minutes
5. a) 2×10^{-4} b) 1.1×10^{-3} c) $6.1 \times 10;^{-3}$

HOMEWORK ACTIVITY: EXAM-STYLE QUESTIONS

1. 4.5×10^5
2. 3.4×10^{-5}
3. 8000 + 20 000 = 28 000 = 2.8×10^4
4. $2 \times 3 \times (10^2 \times 10^3) = 6 \times 10^5$
5. 4×10^{-2}
6. $(6 \div 2) \times (10^5 \div 10^3) = 3 \times 10^2$
7. a) $200 \times 7 \times 10^{-2}$ = 14 mm
b) $500 \times 7 \times 10^{-2}$ = 35 mm = 3.5 cm; $20 \div 3.5$ = 5.7 (1 d.p.) so you would need 6 packs of paper
8. $4.6 \times 10^9 - 2.09 \times 10^5$ = 4 599 800 000 years

GLOSSARY

Standard form
A way of writing very large or very small numbers in the form $A \times 10^n$ where $1 \leq A < 10$ and n is an integer

4 NUMBER: FRACTIONS, DECIMALS AND PERCENTAGES

LEARNING OBJECTIVES

- Convert between fractions, decimals and percentages
- Find a percentage of a number or quantity
- Solve problems involving percentage change, including simple and compound interest and growth and decay and original value problems
- Express one quantity as a percentage of another and calculate percentage change

SPECIFICATION LINKS

- N3, N10, N12, R3, R9, R16

STARTER ACTIVITY

- **What is missing?; 5 minutes; page 36**
  Remind the student that any percentage can be written as a fraction and as a decimal. Model this if necessary.
  Ask the student to identify which term is missing. Each percentage should have an equivalent fraction and decimal.

MAIN ACTIVITIES

- **Percentages and fractions; 25 minutes; page 37**
  Explain how to find percentages and fractions of numbers both mentally and using a multiplier. Discuss how to solve compound interest and original price problems.
- **What percentage or fraction?; 15 minutes; page 38**
  Model how to write one number as a percentage of another by first writing the fraction and then converting to a percentage. Establish that percentage change is calculated using the formula $\dfrac{\text{change}}{\text{original value}} \times 100$.

PLENARY ACTIVITY

- **Describe the method; 5 minutes**
  Ask the student to tell you how they would:
  a) find 20% without a calculator
  b) find 17% with a calculator
  c) calculate the new price of an item if the price increases by 12%
  d) explain the difference between compound interest and simple interest
  e) calculate the percentage change if the original price was £12 and the new price was £15.

HOMEWORK ACTIVITY

- **How hot?; 60 minutes; page 39**
  The student should complete enough questions to collect at least 16 chillies. If they need extra practice with certain types of question, you could select some of the questions for them to answer. You may wish to challenge stronger students to complete as many questions as they can in a certain time.

SUPPORT IDEA

- **Percentages and fractions** Support the student by modelling each question using the 'bar model'.
  For example, for question 8:

  120%

£540

  Discuss how the student could find 1% (divide by 120) and then 100% (× 100).

EXTENSION IDEA

- **Percentages and fractions** For questions 6 and 7, discuss how we could multiply by a power of the multiplier rather than using repetitive multiplication. Ask the student questions such as *how long before the balance in the account is greater than £2,500 in each scenario?* or *how long before the number of insects has halved?*.

PROGRESS AND OBSERVATIONS

STARTER ACTIVITY: WHAT IS MISSING?   TIMING: 5 MINS

LEARNING OBJECTIVES	EQUIPMENT
• Convert between fractions, decimals and percentages	none

1. Each percentage in the cloud should have an equivalent fraction and decimal but one is missing. Give the missing number by grouping the equivalent values.

0.75   66%

$\frac{33}{50}$ $\frac{1}{4}$

0.2 $\frac{1}{2}$

5% $\frac{1}{100}$

0.5

25%

0.66   50%   20%   0.3   10%

30%

1%

0.1   0.05   0.6

0.01 $\frac{1}{10}$

$\frac{1}{5}$ $\frac{6}{10}$

60%   75%   0.25

$\frac{3}{4}$

$\frac{1}{20}$

36

MATHS
— FOUNDATION —

MAIN ACTIVITY: PERCENTAGES AND FRACTIONS   **TIMING: 25 MINS**

LEARNING OBJECTIVES

- Find a percentage of a number
- Use percentages to solve problems
- Interpret fractions and percentages as operators

EQUIPMENT

- calculator

 1. **Work out:**

a) $\frac{1}{5}$ of £25 ....   b) $\frac{2}{3}$ of 81 cm ....   c) $\frac{3}{4}$ of 128 inches ....

 2. Alice says *since you work out 10% by dividing by 10, you must work out 5% by dividing by 5*. Wesley says she is wrong and draws the diagram opposite to prove it.

100%

What percentage is actually worked out by dividing by five?

....

 3. **Complete this table without using a calculator.**

	50%	25%	75%	10%	5%	1%	11%	40%	45%
£320									

 4. **Fill in the missing numbers:**

a) 50% of 320 cm = 0.5 × 320

         = ...

b) 91% of 65 kg = ... × ....

         = ...

c) 18% of £500 = ...× £500

         = ...

d) 20% of 3 hours = ... × ....

         = ... hours

         = ... mins

 5. In a sale, the cost of a coat is reduced by 15%. The coat originally cost £69. Work out the sale price of the coat.

....

 6. A population of an insect colony decays at a rate of 20% per year. One year there are 200 000 insects. How many will be left in 3 years?

....

 7. The price of a TV after 20% VAT has been added is £540. Work out the price of the TV before VAT.

....

MAIN ACTIVITY: WHAT PERCENTAGE OR FRACTION?　　　TIMING: 15 MINS

LEARNING OBJECTIVES

- Express one quantity as a fraction or percentage of another
- Calculate percentage change

EQUIPMENT

- calculator

1. This table shows the number of people using a train one morning.

	Men	Women	Boys	Girls	Babies
Number	10	30	12	8	15

a) What percentage of the passengers were:

　　i)　babies　.............................　　ii)　women?　.............................

b) What fraction of the passengers were:

　　i)　men　.............................　　ii)　girls?　.............................

c) What percentage of the **adults** were men?　...

d) What fraction of the children (not including babies) were girls?　...

e) The following day the number of women using the train was 42. Work out the percentage increase in the number of women.

2. Which formula is used to calculate percentage change? Tick the correct answer.

$$\frac{\text{new value}}{\text{original value}} \times 100$$

$$\frac{\text{change}}{\text{new value}} \times 100$$

$$\frac{\text{original value}}{\text{new value}} \times 100$$

$$\frac{\text{change}}{\text{original value}} \times 100$$

3. Ahmed notices that the cost of his daily newspaper has gone up from 80p to £1. He says *the price has increased by 20% since 20p is 20% of £1.00.* Ahmed is wrong. Explain why.

HOMEWORK ACTIVITY: HOW HOT?

TIMING: 60 MINS

LEARNING OBJECTIVES

- Convert between fractions, decimals and percentages
- Find a percentage of a number or quantity
- Solve problems involving percentage change, including simple interest and original value problems

EQUIPMENT

- calculator

1. There are 16 challenges in the table below. Each row is a different level, each with a different number of chillies. The more chillies, the 'hotter' the challenge.

   You need to collect at least 16 chillies in total. Questions in the first row are worth one chilli, questions in the second row are worth two, and so on. Answer at least one question from each row.

Explain in words how you would work out 50% and 25% of a number without using a calculator.	Write 45% as a fraction.	Write 12% as a decimal.	Write $\frac{15}{20}$ as a percentage.
Work out 35% of 40 cm without using a calculator.	Work out 18% of £120. You may use a calculator, but you must show all your working.	Which is larger, 30% of £15 or 20% of £20?	Reduce 800 by 12%. You may use a calculator, but you must show all your working.
The cost of a t-shirt has increased from £6 to £7.50. What is the percentage change?	Ami invested £250 in a bank account that pays 4% simple interest. How much will she have in the account after 2 years?	In a year, a child grew from 105 cm to 111 cm. Work out the percentage change in height. Give your answer to 2 significant figures.	The surface area of an ice cap decreases by 10% a year. If the area is 2500 m² this year, what will its size be in three years' time?
I think of a number. 40% of that number is 12. What number was I thinking of?	In a sale, all prices are reduced by 25%. If the sale price of a washing machine is £180, what was the original price?	If 7% of a number is 0.21, what was the original number?	Prices in a shop are all increased by 10%. A chocolate bar now costs £0.99. What was the original price?

4 ANSWERS

STARTER ACTIVITY: WHAT IS MISSING?

1. $\dfrac{3}{10}$

MAIN ACTIVITY: PERCENTAGES AND FRACTIONS

1. a) £5   b) 54 cm   c) 96 inches
2. 20%
3.

	50%	25%	75%	10%	5%	1%	11%	40%	45%
£320	£160	£80	£240	£32	£16	£3.20	£35.20	£128	£144

4. a) 160 cm   b) 0.91 × 65 = 59.15 kg   c) 0.18, £90   d) 0.2 × 3 = 0.6 hours = 36 mins
5. £58.65
6. 102 400
7. £450

MAIN ACTIVITY: WHAT PERCENTAGE OR FRACTION?

1. a) i) 20%   ii) 40%   b) i) $\dfrac{2}{15}$ ii) $\dfrac{8}{75}$ c) 25%   d) $\dfrac{2}{5}$ e) 40%

2. $\dfrac{\text{change}}{\text{original value}} \times 100$

3. Ahmed has calculated the increase as a percentage of the new price, not the original price.

HOMEWORK ACTIVITY: HOW HOT?

1.

50%: divide by 2 (or halve it) 25%: divide by 4 (or halve and halve again)	$\dfrac{9}{20}$	0.12	75%
14 cm	0.18 × £120 = £21.60	30% of £15 = £4.50 20% of £20 = £4 30% of £15 is larger.	0.88 × 800 = 704 (accept 0.12 × 800 = 96; 800 − 96 = 704)
25%	£270	5.7%	1822.5 m²
30	£240	3	£0.90

GLOSSARY

Simple interest
Interest calculated on the original investment only, not on any interest earned

Compound interest
Interest calculated on the original investment and any interest earned previously

5 NUMBER: ROUNDING AND ERROR INTERVALS

LEARNING OBJECTIVES

- Round numbers to a given power of 10, the nearest integer, or a given number of decimal places or significant figures
- Use inequality notation to specify simple error intervals
- Apply and interpret limits of accuracy

SPECIFICATION LINKS

- N9, N13, N15, N16

STARTER ACTIVITY

- **Rounding to 100; 5 minutes; page 42**
  Give the student one minute to list as many numbers as they can that round to 100 when rounded to the nearest hundred. Encourage them to include decimals and fractions. At the end, ask them to identify the largest and smallest numbers that round to 100.

MAIN ACTIVITIES

- **Rounding; 20 minutes; page 43**
  Remind the student how to round to the nearest 1000, 100, 10 and 1, displaying the information on a number line if necessary. Establish the 'high five' rule (if it's exactly half way between two values, we round up). Extend this to rounding to a given number of decimal places. Ask the student to answer question 1. Model how to round to a given number of significant figures, then ask the student to answer question 2.

- **Error intervals; 20 minutes; page 44**
  Discuss the answer to question 3 in the first activity and revise how this might be written as an error interval. Explain the difference between ≤ and <, and why we use one of each. Model writing an error interval.

PLENARY ACTIVITY

- **What am I?; 5 minutes**
  Read the student the following and ask them to identify the answer in each case.
  a) I am the largest number with two decimal places that rounds to 3.2. (3.24)
  b) I am the smallest whole number that, when rounded to the nearest 1000, is 5000. (4500)
  c) I am the largest number with one decimal place that is 50 when rounded to 1 significant figure. (54.9)
  If there is time, challenge the student to write their own for you.

HOMEWORK ACTIVITY

- **Rounding and accuracy; 60 minutes; page 45**
  Full instructions are given on the work sheet.

SUPPORT IDEA

- **Rounding** To support rounding, draw out number lines and ask the student to indicate where the number would be on the number line.

EXTENSION IDEA

- **Error intervals** Ask the student to explain a simple way of writing the error interval of any rounded number.

PROGRESS AND OBSERVATIONS

STARTER ACTIVITY: ROUNDING TO 100

TIMING: 5 MINS

LEARNING OBJECTIVES

EQUIPMENT

- Round numbers to a given power of 10

1. In one minute, list as many numbers as you can that round to 100 when rounded to the nearest hundred.

Examples
109, 123.471, 64.7 ...

The smallest number that rounds to 100 is ....

The largest number that rounds to 100 is ....

MAIN ACTIVITY: ROUNDING   TIMING: 20 MINS

LEARNING OBJECTIVES
- Round numbers to a given power of 10, the nearest integer, or a given number of decimal places or significant figures

EQUIPMENT
none

1. **Round each of these numbers to the level of accuracy shown.**

   a) 345 (nearest 10)   ....

   b) 1923 (nearest 100)   ....

   c) 998 (nearest 1000)   ....

   d) 7.4 (nearest whole number) ....

   e) 85.6 (nearest 10)   ....

   f) 43.67 (1 decimal place) ....

   g) 18.256 (2 decimal places) ....

   h) 19.95 (1 decimal place)   ....

2. **Complete this table by rounding each number to 1, 2, 3 or 4 significant figures.**
   The first row has been done for you.

Number	1 significant figure	2 significant figures	3 significant figures	4 significant figures
43 891	40 000	44 000	43 900	43 890
5392				
198 500				
18.3921				
0.0019257				
92 340 000				
0.00071				

3. **The length of a room rounded to 1 significant figure is 2 metres. What is:**

   a) the shortest length it could be?   ....

   b) the longest length it could be?   ....

| MAIN ACTIVITY: ERROR INTERVALS | TIMING: 20 MINS |

LEARNING OBJECTIVES

EQUIPMENT

- Use inequality notation to specify simple error intervals
- Apply and interpret limits of accuracy

- calculator

1. Alex says the length of his run is 2 miles to the nearest mile.

   a) What is the shortest distance he could have run?   ....

   b) What is the furthest distance he could have run?   ....

   c) Write the error interval for the length of his run.   ....

2. These numbers have been rounded to 1 decimal place. Write down an error interval for each one.

   a) 2.3   ....

   b) 9.1   ....

   c) 180.5   ....

3. A builder estimates the cost of a job to be £1800 to the nearest hundred pounds. Write an error interval for the cost of the job (C).

   ....

4. Mr Axby measures the dimensions of a room to the nearest 10 cm. He writes down that the width is 2.6 m and the length is 5.2 m. Carpet costs £12.99 per square metre.

   a) Work out the maximum cost of carpeting the room.

   ....

   b) Work out the minimum cost of carpeting the room.

   ....

HOMEWORK ACTIVITY: ROUNDING AND ACCURACY

TIMING: 60 MINS

LEARNING OBJECTIVES
- Use inequality notation to specify simple error intervals
- Apply and interpret limits of accuracy

EQUIPMENT
- calculator

1. Round each of the numbers to the degree of accuracy shown.

a) 0.379 to 1 decimal place

....

b) 43 920 to 1 significant figure

....

c) 0.2089 to 2 significant figures

....

d) 17.92 to the nearest 10

....

e) 1923.54 to the nearest whole number

....

f) 18.91 to the nearest 100

....

g) 2.0095 to 3 decimal places

....

2. Erica rounds an answer, y, to 1 decimal place. The result is 9.7. Write an error interval for y.

....

3. In a test, students are given this question: 'A number, x, rounded to 3 significant figures is 17.8. Write an error interval for x.' Here are three students' answers:

Johanna: $17.80 \leq x < 17.85$ **Alison: $17.75 < x < 17.85$**   **Romesh: $17.75 \leq x \leq 17.85$**

a) Explain what each student has done wrong.

....

....

b) Give the correct answer.

....

4. Look at this number: 36 789 024.

a) Round the number to 2 significant figures.

....

b) Write your answer to a) in standard form.

....

5. A scientist measures the atomic radius (x) of lithium as 1.45×10^{-10} to three significant figures. Write an error interval for the atomic radius (x).

....

5 ANSWERS

STARTER ACTIVITY: ROUNDING TO 100

1. Any values that round to 100.

The smallest possible number is 50; the largest is $149.99\dot{9}$.

MAIN ACTIVITY: ROUNDING

1. a) 350   b) 1900   c) 1000   d) 7   e) 90   f) 43.7   g) 18.26   h) 20.0

2.

Number	1 significant figure	2 significant figures	3 significant figures	4 significant figures
5392	5000	5400	5390	5392
198 500	200 000	200 000	199 000	198 500
18.3921	20	18	18.4	18.39
0.0019257	0.002	0.0019	0.00193	0.001926
92 340 000	90 000 000	92 000 000	92 300 000	92 340 000
0.00071	0.0007	0.00071	N/A	N/A

3. a) 1.5 m   b) $2.4\dot{9}$ m

MAIN ACTIVITY: ERROR INTERVALS

1. a) 1.5 miles   b) $2.4\dot{9}$ miles   c) $1.5 \leq x < 2.5$
2. a) $2.25 \leq x < 2.35$ b) $9.05 \leq x < 9.15$ c) $180.45 \leq x < 180.55$
3. $1750 \leq C < 1850$
4. a) maximum area = 13.9125 m$_2$ maximum cost = £180.72   b) minimum area = 13.1325 m$_2$ minimum cost = £170.59

HOMEWORK ACTIVITY: ROUNDING AND ACCURACY

1. a) 0.4   b) 40 000   c) 0.21   d) 20   e) 1924   f) 0   g) 2.010
2. $9.65 \leq y < 9.75$
3. a) Johanna has not correctly identified the lower limit.
Alison should have written a $\leq$ sign after 17.75.
Romesh should have written a $<$ sign before 17.85.
b) $17.75 \leq x < 17.85$
4. a) 37 000 000   b) 3.7×10^7
5. $1.445 \times 10^{-10} \leq x < 1.455 \times 10^{-10}$

GLOSSARY

Error interval
The range of values in which a precise value could be

6 NUMBER: ESTIMATING

LEARNING OBJECTIVES

- Apply the four operations, with and without using formal written methods
- Estimate answers
- Check calculations using approximation and estimation, including answers obtained using a calculator
- Recognise and use relationships between operations, including inverse operations

SPECIFICATION LINKS

- N2, N3, N12, N14

STARTER ACTIVITY

- **What do we know?; 5 minutes; page 48**
  In the starter, the student is given a long multiplication calculation and asked to write down several facts based on the calculation. You may wish to remind students of the process of long multiplication before attempting this starter.

MAIN ACTIVITIES

- **Estimations; 20 minutes; page 49**
  Invite the student to suggest how they might find the approximate answer to each of the calculations in the first activity. Encourage them to round the values to the nearest whole number, and to use their knowledge of key number facts to support their reasoning. The student should explain their methods to you as they work.
  This exercise requires use of the priority of operations and simple substitution, so you may wish to revise these ideas first. After completing the matching activity, the student should find the incorrect value using a calculator.

- **Making it real; 20 minutes; page 50**
  Remind the student how to carry out the four operations involving decimals. Emphasise the importance of lining up the decimal point when carrying out addition and subtraction, and reinforce how we deal with the decimal point when carrying out multiplication and division involving decimals. Discuss what is meant by inverse operations and how calculations can be checked using inverse operations. Ask the student to complete questions 1 and 2 without using a calculator.

PLENARY ACTIVITY

- **How much to carpet the room?; 5 minutes**
  Ask the student to estimate the area of the room you are in. Give the cost of underlay (£3.99 per square metre) and carpet (£18.99 per square metre) and ask the student to work out the cost of laying carpet in the room. Is this an underestimate or an overestimate? Ask the student to justify their answer.

HOMEWORK ACTIVITY

- **Estimation and exact solutions; 25 minutes; page 51**
  The student should answer these exam-style questions using the skills covered in this lesson and in the previous lesson.

SUPPORT IDEAS

- **Estimations** Spend some time developing the student's understanding of place value. For example, if 3 × 2 = 6, what must 0.3 × 2 be?
- **Making it real** Encourage the student to use a highlighter to mark the key information in problem-solving questions.

EXTENSION IDEA

- **Making it real** Encourage the student to compare estimates with the exact solution and ask them to work out the percentage error.

PROGRESS AND OBSERVATIONS

STARTER ACTIVITY: WHAT DO WE KNOW?   TIMING: 5 MINS

LEARNING OBJECTIVES

- Recognise and use relationships between operations, including inverse operations

EQUIPMENT

Erica works out 237 × 45 using the long multiplication method.

```
        2   3   7
    ×       4   5
    1   1   8   5
            1   3
    9   4   8   0
        1   2
1   0   6   6   5
            1
```

1. Each of the following calculations can be worked out without doing any more written calculations. Explain to your tutor how you could work them out.

   a) 237 × 5

   b) 237 × 40

   c) 10665 − 9480

   d) 10665 − 1185

   e) 237 × 450

MATHS
— FOUNDATION —

MAIN ACTIVITY: ESTIMATIONS	TIMING: **20** MINS

LEARNING OBJECTIVES
- Estimate answers
- Check calculations using approximation and estimation, including answers obtained using a calculator

EQUIPMENT
- calculator

For these activities, use the following values.

A = 3.04	B = 2.78	C = 94.82	D = 0.405	E = 121.02

1. Harry has worked out the numerical values of the expressions below. Unfortunately, he has muddled his answers up. Explain to your tutor how you would work out the approximate value of each expression, and then draw lines to match each expression with the most likely answer.

2A	363.87
C − B	62.03
2D + 3E	6.08
2(A + E)	7.506
4A − D	9.542
A^2	92.04
$\dfrac{A}{D}$	266.6396
$\dfrac{A+E}{2}$	248.12
A + B × C	11.755

2. One of the calculations is incorrect. Use a calculator to work out which answer is wrong, and what the correct answer should be.

....

MAIN ACTIVITY: MAKING IT REAL   TIMING: 20 MINS

LEARNING OBJECTIVES

- Apply the four operations, with and without using formal written methods
- Recognise and use relationships between operations, including inverse operations
- Estimate answers

EQUIPMENT

- calculator

The prices per person for tickets in different parts of a theatre are shown in the table below:

Position	Stalls	Royal circle	Dress circle	Balcony	Private box (6 seats)
Price	£94.75	£52.65	£48.60	£29.75	£352.14

1. **To work out the cost of two tickets in the stalls and two tickets in the royal circle, Alice calculates 90 + 90 + 50 + 50 = £280.**

   a) Write down how Alice got to this calculation.

   ....

   ....

   b) Is this an overestimate or an underestimate? Write down how you know.

   ....

   ....

2. **Matt has £310 to spend. He wants to buy 6 seats together. Discuss with your tutor how you could work out where he could buy seats. Work out how much change he will have if he buys the most expensive seats he can afford.**

   ....

3. **In order to heat the theatre one night, 2341 units of electricity were used. Each unit costs 11.459p.**

   a) Estimate the cost of heating for that night.

   ....

   b) Use a calculator to work out the cost of heating the theatre. Round your answer to the nearest penny.

   ....

   c) On a different night, the cost of heating the theatre is £228.65 to the nearest penny. Look at the list below and tick the calculation you would use to work out how many units were used.

   A: 228.65 × 11.459   B: 228.65 − 11.459   C: 11.459 ÷ 228.65

   D: 22865 ÷ 11.459   E: 228.65 ÷ 11.459

HOMEWORK ACTIVITY: ESTIMATION AND EXACT SOLUTIONS TIMING: 25 MINS

LEARNING OBJECTIVES

- Apply the four operations, with and without using formal written methods
- Estimate answers
- Check calculations using approximation and estimation, including answers obtained using a calculator

EQUIPMENT

- calculator

1. **Monica is considering purchasing a new mobile phone contract. Each tariff requires you to sign up for a year.**

   Tariff A: 5p per minute of calls, 2p per text

   Tariff B: Monthly line rental of £9.99, 120 minutes free calls then 4p per minute, 100 free texts then 5p per text

   Monica uses an average of 200 minutes and sends 60 texts each month.
   Use this information to work out which tariff Monica should choose.

   ----

2. **Look at this calculation:** $\dfrac{3.09 \times 19.5 + 23.2}{99.2 - 0.25}$

   a) Give an approximate answer by rounding each number to 1 significant figure.

   ----

   b) Work out the exact value. Write down all the figures on your calculator.

   ----

3. **A sports field measures 519 m by 205 m. Grass seed costs £19.99 per box and each box covers an area of 250 m².**

   a) Work out an estimate for the cost of grass seed to cover the sports field.

   ----

   b) Is your answer an overestimate or an underestimate? Give a reason for your answer.

   ----

6 ANSWERS

STARTER ACTIVITY: WHAT DO WE KNOW?

1. a) $237 \times 5 = 1185$ (the first line of long multiplication)

b) $237 \times 40 = 9480$ (the second line of long multiplication)

c) $10\,665 - 9480 = 1185$ (answer – second line of calculation)

d) $10\,665 - 1185 = 9480$ (answer – first line of calculation)

e) $237 \times 450 = 106\,650$ (answer × 10)

MAIN ACTIVITY: ESTIMATIONS

1. $2A = 6.08$ $C - B = 92.04$ $2D + 3E = 363.87$ $2(A + E) = 248.12$ $4A - D = 11.755$

$A^2 = 9.542$ $\dfrac{A}{D} = 7.506$ $\dfrac{A + E}{2} = 62.03$ $A + B \times C = 266.6396$

2. The answer to A^2 is incorrect – it should be 9.2416.

MAIN ACTIVITY: MAKING IT REAL

1. a) She has rounded the prices to 1 significant figure and then added them.

b) It is an underestimate as all the prices were rounded down.

2. Matt could round the price for a seat and multiply it by 6, or divide £310 by 6 and find the closest price to this.

The most expensive seats Matt can afford are in the dress circle. He will have £18.40 change.

3. a) $10p \times 2300 = 23\,000p$

$= £230.00$

b) £268.26

c) D

HOMEWORK ACTIVITY: ESTIMATION AND EXACT SOLUTIONS

1. Tariff A: texts cost $60 \times 2 = 120 = £1.20$ calls cost $200 \times £0.05 = £10.00$ total cost = £11.20

Tariff B: texts are all free   calls cost $80 \times 4 = 320 = £3.20$ total cost = £9.99 + £3.20 = £13.19

Tariff A is better for Monica.

2. a) 0.8   b) 0.8434057605

3. a) $500 \times 200 = 100\,000$ $100\,000 \div 250 = 400$ $400 \times £20 = £8000$

b) The answer is an underestimate as the lengths have both been rounded down.

7 ALGEBRA: STRAIGHT LINE GRAPHS

LEARNING OBJECTIVES

- Plot graphs of equations that correspond to straight line graphs in the coordinate plane
- Use the form $y = mx + c$ to identify parallel lines
- Find the equation of a straight line given two points or one point and the gradient
- Interpret graphs in real contexts

SPECIFICATION LINKS

- A8, A9, A14

STARTER ACTIVITY

- **Match it up!; 5 minutes; page 54**
  Ask the student to match the graphs to their equations ($y = \pm a$, $x = \pm a$, $y = x$ and $y = -x$).

MAIN ACTIVITIES

- **Plotting graphs; 15 minutes; page 55**
  Model how to plot a graph of the form $y = mx + c$. When they have drawn the graphs, help the student to recognise the relationship between the gradient and the m value in the equation. Encourage them to notice that for an increase of 1 in the x-direction the y value will increase by 'm'. Explain that the c value gives the y-intercept. Notice the relationship between the equation of the graph and the gradient and y-intercept. Ask the student what they notice about graphs A and D.

- **Working with graphs; 25 minutes; page 56**
  Encourage the student to think about what sort of data they have seen represented on graphs, and why they are a useful way of displaying information. Establish that the gradient of a graph is its rate of change. Work through question 1. For question 2, remind the student that to find the gradient and y-intercept from a graph's equation, it must be in the form $y = mx + c$. If necessary, model rearranging an equation given in a different form. During question 3, discuss anything that arises from the points the student chooses to plot (e.g. a negative or fractional gradient). Model how to find the equation of a straight line given a point and the gradient by using substitution to find c.

PLENARY ACTIVITY

- **Speed–time graph; 5 minutes**
  Explain that you are going to draw a speed–time graph of a car on a 10 mile journey. Sketch the graph with the student, discussing the initial speed, final speed, acceleration and deceleration as you do so.

HOMEWORK ACTIVITY

- **Exam-style questions; 45 minutes; page 57**
  The student should work independently through the exam-style questions. You may wish to provide axes for the solutions to be drawn onto for less able students.

SUPPORT IDEA

- **Plotting graphs** Encourage the student to explain in words the relationship between the x and y coordinates.

EXTENSION IDEA

- **Working with graphs** Challenge the student to decide whether a pair of straight line graphs will intercept and to justify how they know. Discuss with the student how to find the gradient of a line given two points, without drawing the graph.

PROGRESS AND OBSERVATIONS

STARTER ACTIVITY: MATCH IT UP!

TIMING: 5 MINS

LEARNING OBJECTIVES	EQUIPMENT
• Find the equation of a straight line given two points or one point and the gradient	none

1. **Match each equation to the correct graph by writing the appropriate letter next to each equation.**

$y = 3$ .... $x = 3$ .... $x = -2$ .... $y = 2$ ....

$x = 0.5$ .... $y = -3$ .... $y = x$ .... $y = -x$ ....

A
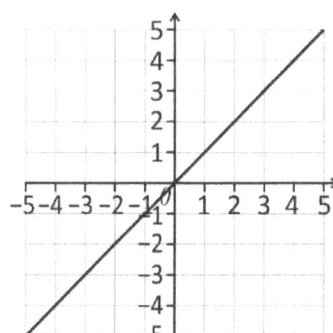

B
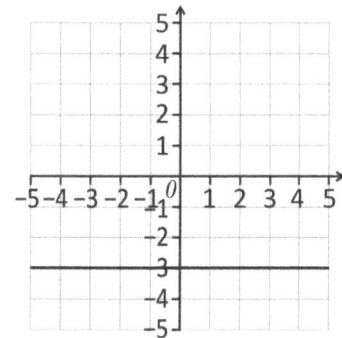

C
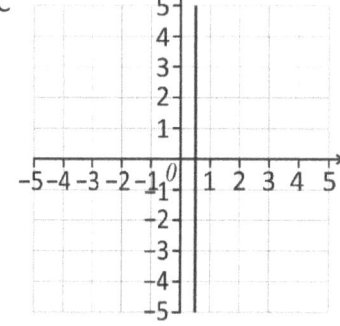

D
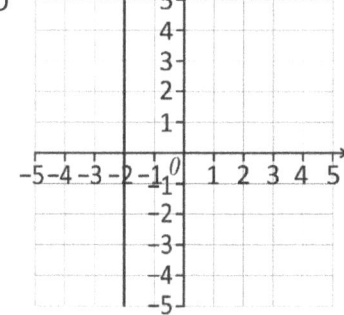

E
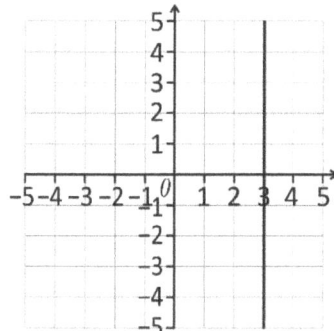

F
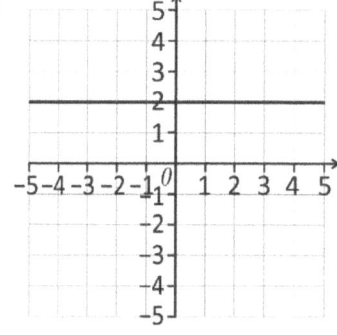

G
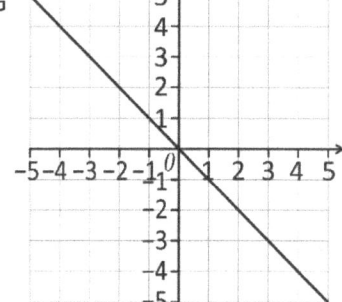

H
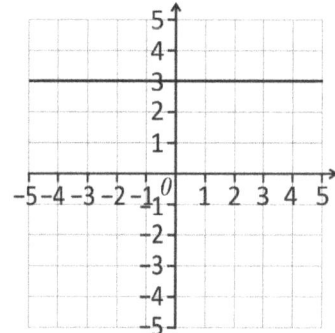

MATHS
— FOUNDATION —

MAIN ACTIVITY: PLOTTING GRAPHS   **TIMING: 15 MINS**

LEARNING OBJECTIVES

- Plot graphs of equations that correspond to straight line graphs in the coordinate plane
- Use the form $y = mx + c$ to identify parallel lines

EQUIPMENT

- ruler
- graph paper

 1. **For each equation, complete the tables of values and plot the graph on the axes provided.**

a) $y = 5x + 2$

x	–5	–4	–3	–2	–1	0	1	2	3	4	5
y											

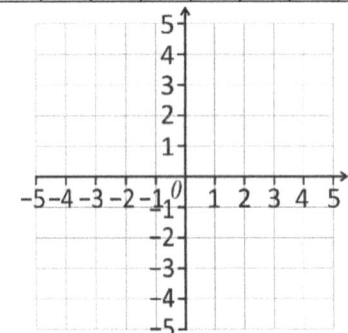

b) $y = 2x - 1$

x	0	–5	–4	–3	–2	–1	0	1	2	3	4	5
y												

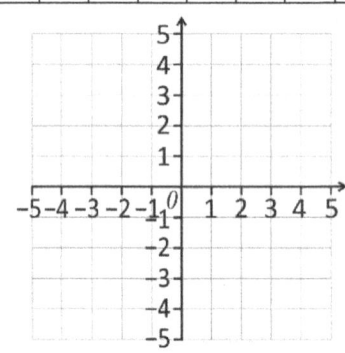

c) $y = x + 3$

x	–5	–4	–3	–2	–1	0	1	2	3	4	5
y											

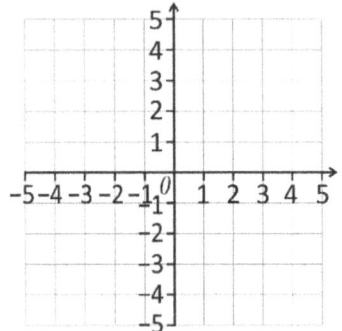

d) $y = 5x - 4$

x	0	–5	–4	–3	–2	–1	1	2	3	4	5
y											

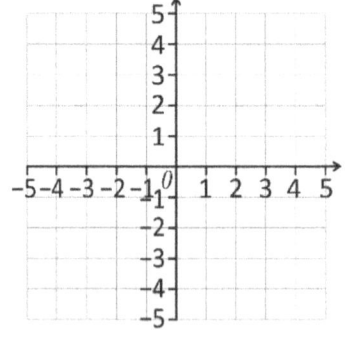

 2. **For each graph, work out the gradient and the y-intercept.**

a) gradient = ...   y-intercept = ...

b) gradient = ...   y-intercept = ...

c) gradient = ...   y-intercept = ...

d) gradient = ...   y-intercept = ...

MAIN ACTIVITY: WORKING WITH GRAPHS

TIMING: 25 MINS

LEARNING OBJECTIVES

- Use the form $y = mx + c$ to identify parallel lines
- Find the equation of a straight line given two points or one point and the gradient
- Interpret graphs in real contexts

EQUIPMENT

- ruler

1. **The graph shows the exchange rate between Emirati dirham (AED) and pounds (£).**

   a) How much is £1 worth in AED?

   ....

   b) How much is 1 AED worth in pounds?

   ....

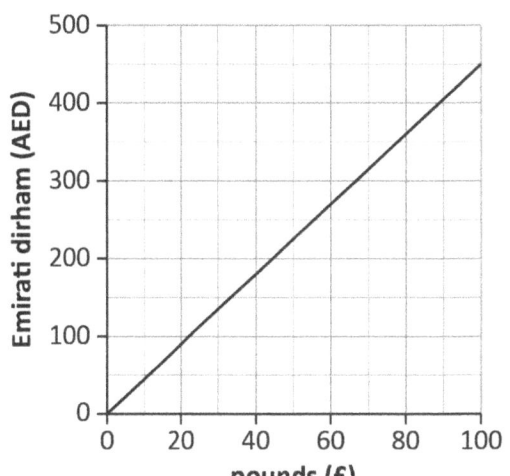

2. **Here are the equations of four straight line graphs:**

   graph A: $y = 3x - 5$ graph B: $3y + 6x = -15$ graph C: $x - y = 10$ graph D: $3x = 3y - 2$

   a) Which two graphs are parallel?

   b) Which two graphs intersect the y-axis at the same point? ....

   c) Explain to your tutor how you worked out your answers.

3. **Mark any two points on this coordinate axis and use a ruler to draw a line between them.**

   a) Calculate the gradient.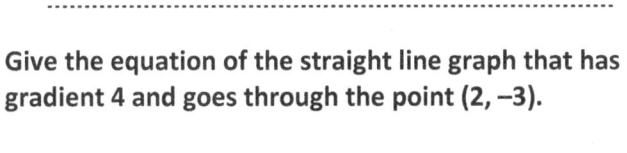

   b) Give the equation of the line joining the two points.

   ....

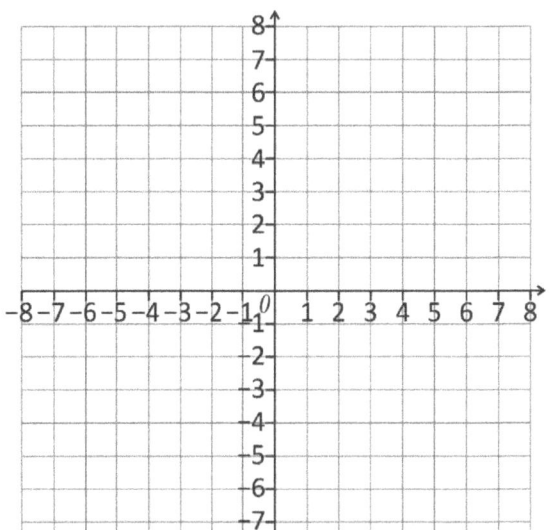

4. **Give the equation of the straight line graph that has gradient 4 and goes through the point (2, –3).**

   ....

HOMEWORK: EXAM-STYLE QUESTIONS

TIMING: 45 MINS

LEARNING OBJECTIVES

- Plot graphs of equations that correspond to straight line graphs in the coordinate plane
- Use the form $y = mx + c$ to identify parallel lines
- Find the equation of a straight line given two points or one point and the gradient
- Interpret graphs in real contexts

EQUIPMENT

- ruler
- graph paper

1. **On graph paper, draw a set of axes with −10 ≤ x ≤ 10 and −10 ≤ y ≤ 10.**

    a) On the axes, draw the graphs of $y = x$, $x = -2$ and $y = -3$.

    b) Write down the coordinates of the three points where these lines intersect to create a triangle.

----

2. **a) Write down the equation of a straight line with gradient 4 and y-intercept 8.**

----

    b) Which of these straight line graphs is parallel to the graph in part a)? Circle the relevant equation.

    A: $2x - 4y = 12$ B: $2y - 8x = -6$ C: $y + 4x = 0$

3. **Two points have been plotted on the axes opposite.**

    a) Write down the coordinates of the point A.   ....

    b) Work out the midpoint of the line AB.   ....

    c) Work out the gradient of the line AB.   ....

    d) Write down the equation of the line AB.   ....

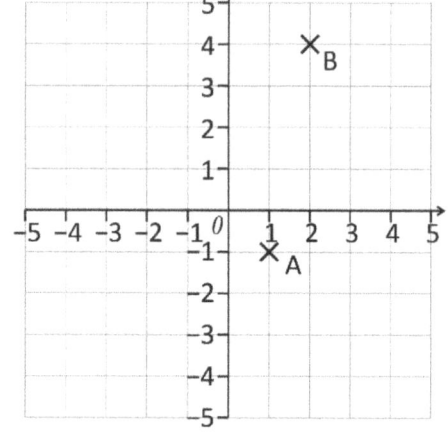

4. **5 miles = 8 km**

    a) Convert 25 miles into kilometres.   ....

    b) On graph paper, draw a set of axes with 0 ≤ x ≤ 50, and 0 ≤ y ≤ 80. Label the x-axis 'miles' and the y-axis 'kilometres'. Draw a graph to convert between miles and kilometres.

    c) Use your graph to convert 52 km into miles. Show clearly where you take your readings.

----

7 ANSWERS

STARTER ACTIVITY: MATCH IT UP!

1. A: $y = x$ B: $y = -3$ C: $x = 0.5$ D: $x = -2$ E: $x = 3$ F: $y = 2$ G: $y = -x$ H: $y = 3$

MAIN ACTIVITY: PLOTTING GRAPHS

1. Check student's answers.
2. a) gradient = 5, y-intercept = 2
b) gradient = 2, y-intercept = −1
c) gradient = 1, y-intercept = 3
d) gradient = 5, y-intercept = −4
If the graph is in the form $y = mx + c$, m = gradient, c = y-intercept.

MAIN ACTIVITY: WORKING WITH GRAPHS

1. a) £1 ≈ 4.5 AED   b) 1 AED ≈ £0.22
2. a) C and D   b) A and B   c) Student's own answers
3. Student's own answers
4. $y = 4x - 11$

HOMEWORK: EXAM-STYLE QUESTIONS

1. a) Check student's graph.   b) (−2, −2)   (−3, −3)   (−2, −3)
2. a) $y = 4x + 8$ b) B
3. a) (1, −1)   b) (1.5, 1.5)   c) 5   d) $y = 5x - 6$
4. a) 40 km
b)

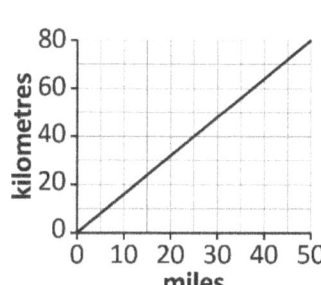

c) 32.5 miles – check student's measurement on their graph.

GLOSSARY

Gradient

The steepness of a line, measured as change in y over change in x (or $\dfrac{\text{distance up}}{\text{distance across}}$)

8 ALGEBRA: QUADRATIC, CUBIC AND RECIPROCAL GRAPHS

LEARNING OBJECTIVES

- Recognise, sketch and interpret graphs of quadratic functions, simple cubic functions and the reciprocal function $y = \dfrac{1}{x}$

SPECIFICATION LINKS

- A8, A11, A12

STARTER ACTIVITY

- **Which equation?; 5 minutes; page 60**
  Match the equations of linear graphs to their corresponding graphical representation.

MAIN ACTIVITIES

- **Plotting graphs; 20 minutes; page 61**
  Discuss the shape of a quadratic function and explain that you can plot a graph of one by first completing a table of values. Ask the student to complete question 1. Explain that when the highest power of x is 3, we call the graph a cubic graph. Ask the student to complete question 2, discussing any properties that the quadratic and cubic graphs have in common.
- **Properties of graphs; 20 minutes; page 62**
  Define the terms 'root', 'intercept', 'turning point', 'maximum' and 'minimum'. Make sure the student knows what a turning point looks like on a cubic graph, as well as on a quadratic graph. Emphasise that they may need to find approximate values from graphs.

PLENARY ACTIVITY

- **What shape am I?; 5 minutes**
  Write down the equation of some quadratic, cubic and reciprocal graphs. Ask the student to sketch the shape of the graph and explain which values could be found from a graphical representation of the graph (minimum, maximum, roots, intercept and where they would be found on the graph).

HOMEWORK ACTIVITY

- **Interpreting graphs; 60 minutes; page 63**
  Full instructions are given on the activity sheet.

SUPPORT IDEA

- **Plotting graphs** If necessary, remind the student how to substitute values into algebraic expressions, modelling the first couple of calculations.

EXTENSION IDEAS

- **Properties of graphs** Ask the student to describe the properties of quadratic, cubic and reciprocal graphs, e.g. as the x-value increases, the y-value decreases. Encourage them to consider how the gradient of each graph changes.
- **What shape am I?** Give the student enough details (e.g. the y-intercept and a gradient or the x- and y-intercepts) and ask them to sketch a quadratic function.

PROGRESS AND OBSERVATIONS

Starter activity: Which equation?   Timing: 5 mins

Learning objectives
- Recognise graphs of linear functions

Equipment
none

1. Eight graphs have been drawn on the coordinate axis. Next to each equation, write the letter of the line on the graph that represents it. Explain to your tutor how you can work out which graph matches which equation.

$y = 2x$ .... $y = 4$ ....

$y = -3x$ .... $x = -5$ ....

$y = 3x + 4$ .... $y = -2x - 2$ ....

$y = -x$ .... $y = 5x + 2$ ....

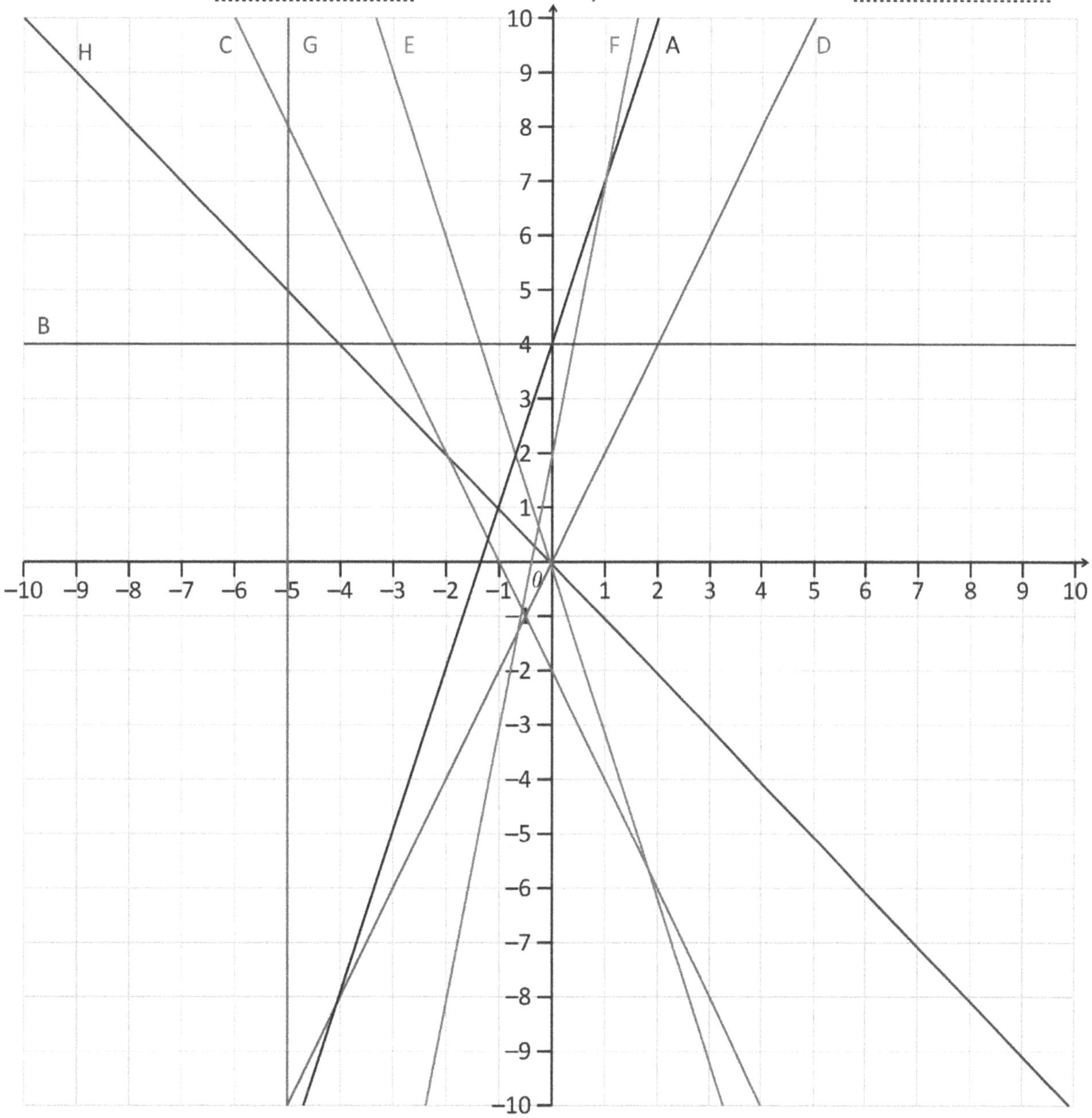

MATHS
— FOUNDATION —

| MAIN ACTIVITY: PLOTTING GRAPHS | TIMING: 20 MINS |

LEARNING OBJECTIVES

- Plot graphs of quadratic functions, simple cubic functions and the reciprocal function $y = \dfrac{1}{x}$

EQUIPMENT

- graph paper
- ruler

 1. Complete the table of values for the quadratic graph $y = x^2 + 2x - 3$, then plot the graph on a suitable set of axes.

x	−3	−2	−1	0	1	2	3
x^2	9				1		
2x	−6				2		
−3	−3				−3		
y	0				0		

 2. This table should show the values for the cubic graph $y = x^3$.

    a) Complete the table and plot the graph on a suitable set of axes.

x	−3	−2	−1	0	1	2	3
y							

    b) Describe what you notice about the shape of quadratic and cubic graphs.

Remember, the highest power of x in a quadratic graph is x^2. The highest power of x in a cubic graph is x^3.

 3. This table should show the values for $y = \dfrac{1}{x}$.

    a) Complete the table and plot the graph on a suitable set of axes.

x	0.1	0.2	0.5	0.8	1	2	4	5
y								

    b) Describe what you notice about this graph.

MAIN ACTIVITY: PROPERTIES OF GRAPHS   TIMING: 20 MINS

LEARNING OBJECTIVES

- Interpret graphs of quadratic functions, simple cubic functions and the reciprocal function $y = \dfrac{1}{x}$

EQUIPMENT

1. **Match each word to its definition.**

root	The smallest value that an equation can have for any value of the variable
y-intercept	An equation with the highest power of the variable being 3
turning point	The point where a graph crosses the *x*-axis, where *y* = 0
minimum	The largest value that an equation can have for any value of the variable
maximum	The point where the graph's gradient is 0; it can be a maximum or minimum point, or a point of inflection
quadratic equation	An equation with the highest power of the variable being 2
cubic equation	The point where a graph crosses the *y*-axis, where *x* = 0

2. **This is the graph of $y = x^2 + 2x - 3$.**

   Discuss with your tutor how you would give:

   a) the coordinates where the graph crosses:

      i)   the *x*-axis

      ii) the *y*-axis

   b) the coordinates of the turning point of the graph

   c) the equation of the line of symmetry of the graph.

3. **For each of the graphs in the *Plotting graphs* activity, work out:**

   a) any turning points

   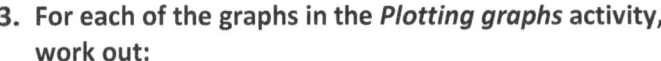

   b) the *y*-intercept (if applicable)

   c) any roots

   d) the equation of any lines of symmetry.

HOMEWORK ACTIVITY: INTERPRETING GRAPHS

TIMING: 60 MINS

LEARNING OBJECTIVES
- Recognise linear, quadratic, cubic and reciprocal graphs
- Interpret graphs of quadratic functions, simple cubic functions and the reciprocal function $y = \dfrac{1}{x}$

EQUIPMENT
- graph paper
- ruler
- coloured pencils

1. **a) Look at the shape of each graph and decide if it is linear, quadratic, cubic or reciprocal.**

i)
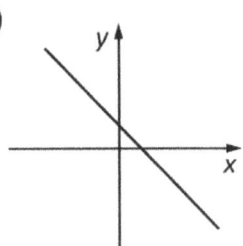

ii)
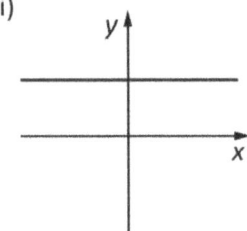

iii)
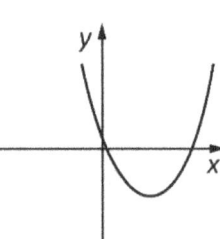

iv)
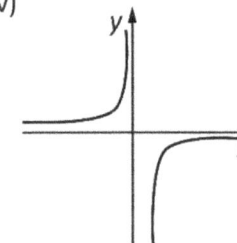

v)
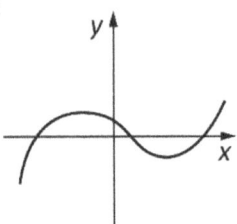

vi)
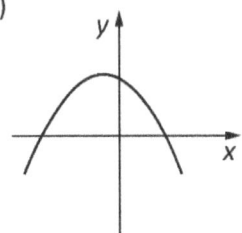

vii)
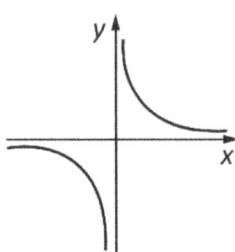

viii)
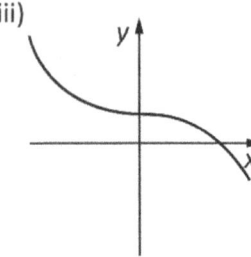

i) .... ii) .... iii) .... iv) ....

v) .... vi) .... vii) .... viii) ....

b) Mark any turning points with a blue cross.
c) Mark any roots with a green cross.
d) Mark any points of intersection with the *y*-axis with a red cross.
e) Draw a dotted line to indicate any lines of symmetry.

2. **For the graph of *y* = 2*x*² + 3*x* − 5, give:**

   a) the value of *y* when *x* = 0 ....

   b) the values of *x* when *y* = 0 ....

   c) the smallest value of *y* ....

   d) the equation of the line of symmetry ....

   e) Explain why you can't give the largest value of *y*.

   ....

3. **Plot these graphs on graph paper and label as many of their properties as you can.**

   a) $y = \dfrac{4}{x}$

   b) $y = x^3 + 5$

   c) $y = 3x^3 - 2x^2 + 1$

8 ANSWERS

STARTER ACTIVITY: WHICH EQUATION?

1. A: $y = 3x + 4$ B: $y = 4$ C: $y = -2x - 2$ D: $y = 2x$
E: $y = -3x$ F: $y = 5x + 2$ G: $x = -5$ H: $y = -x$

Student's explanations should refer to the lines' gradients and intercepts.

MAIN ACTIVITY: PLOTTING GRAPHS

1. Check student's graph.

x	−3	−2	1	0	1	2	3
y	0	−3	−4	−3	0	5	12

2. a) Check student's graph.

x	−3	−2	1	0	1	2	3
y	−27	−8	−1	0	1	8	27

b) Quadratic graphs are a U-shape (a parabola). Cubic graphs are an elongated, sideways S-shape.

3. a)

x	0.1	0.2	0.5	0.8	1	2	4	5
y	10	5	2	1.25	1	0.5	0.25	0.2

b) As the x-value increases, the y-value decreases.

MAIN ACTIVITY: PROPERTIES OF GRAPHS

1. root: The point where a graph crosses the x-axis, where $y = 0$

y-intercept: The point where a graph crosses the y-axis, where $x = 0$

turning point: The point where the graph's gradient is 0; it can be a maximum or minimum point, or point of inflection

minimum: The smallest value that an equation can have for any value of the variable

maximum: The largest value that an equation can have for any value of the variable

quadratic equation: An equation with the highest power of the variable being 2

cubic equation: An equation with the highest power of the variable being 3

2. a) i) (−3, 0) and (1, 0) ii) (0, −3) b) (−1, −4) c) $x = -1$

3. graph 1: a) (−1, −4) b) (0, −3) c) (−3, 0) and (1, 0) d) $x = -1$

graph 2: a) (0, 0) b) (0, 0) c) (0, 0) d) N/A

graph 3: a) N/A b) N/A c) N/A d) $y = x$

HOMEWORK ACTIVITY: INTERPRETING GRAPHS

1. a) i) linear ii) linear iii) quadratic iv) reciprocal
v) cubic vi) quadratic vii) reciprocal vii) cubic

b)–e) Check student's graphs.

2. a) −5 b) 1 and −2.5 c) approximately −6 d) $x = -0.75$

e) The value of y will continue to increase as x increases or decreases.

3. Check student's graphs.

GLOSSARY

Turning point

A point at which the gradient of a graph is zero

Root

The value when a function is equal to zero

Maximum

A point at which the gradient of a graph changes from positive to negative

Minimum

A point at which the gradient of a graph changes from negative to positive

9 ALGEBRA: INTERPRETING GRAPHS

LEARNING OBJECTIVES

- Draw and interpret distance–time graphs
- Identify and interpret roots, intercepts, turning points and gradients of graphs, relating them to real-life problems

SPECIFICATION LINKS

- A10, A11, A12, A14

STARTER ACTIVITY

- **Filling vases; 5 minutes; page 66**
  Explain that water is poured into each vase at a steady rate. Ask the student to match each graph showing height of water against time with the correctly shaped vase. Discuss with the student how the gradient of the graph changes over time for each vase.

MAIN ACTIVITIES

- **Interpreting graphs; 15 minutes; page 67**
  Discuss with the student how to sketch a linear graph using the *y*-intercept and gradient. Use the discussion to elicit the idea that gradient represents rate of change.
- **Distance–time and speed–time graphs; 25 minutes; page 68**
  Discuss the difference between distance–time and speed–time graphs. Establish that the speed can be found from a distance–time graph by calculating the gradient, remind students how to do this if necessary.

PLENARY ACTIVITY

- **What does the gradient tell us?; 5 minutes**
  Discuss how to find the gradient of a graph (change in *y* over change in *x*) and establish what the gradient means on a distance–time graph. Extend this to understanding rate of change and how the gradient describes the change in the *y*-direction for a change of one in the *x*-direction. Explain that the gradient can sometimes be interpreted in the context of 'real-life' situations; refer to the starter activity and question 2 in the 'Interpreting graphs' activity.

HOMEWORK ACTIVITY

- **All types of graph; 60 minutes; page 69**
  Full instructions are given on the activity sheet.

SUPPORT IDEA

- **Distance–time and speed–time graphs** Simplify the problem by drawing distance–time and speed–time graphs which are linear. Discuss what the gradient shows in each. Extend to what a horizontal line would mean on each type of graph.

EXTENSION IDEA

- **Distance–time and speed–time graphs** Invite the student to draw a distance–time graph of a car accelerating from rest. Discuss the changes in gradient and why they occur (acceleration) and what shape we would expect the graph to be.

PROGRESS AND OBSERVATIONS

MATHS
— FOUNDATION —

AQA

STARTER ACTIVITY: FILLING VASES

TIMING: 5 MINS

LEARNING OBJECTIVES

- Relate the gradient of graphs to real-life situations

EQUIPMENT

Five different vases are filled with water at a steady rate. The graphs show the height of water in each vase over time. Discuss with your tutor which graph matches which vase. Make sure you can give a reason.

1.

A.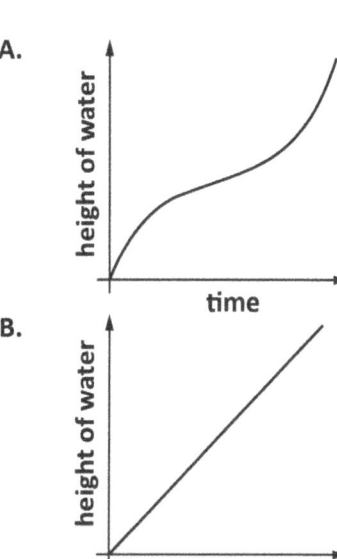

2.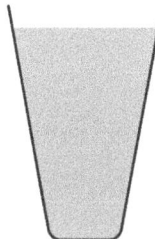

B.

3.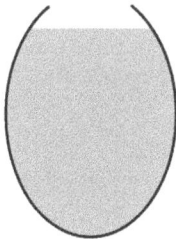

C.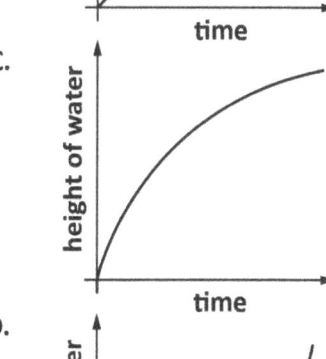

4.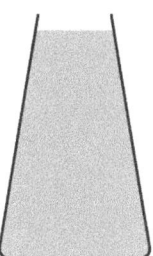

D.

5.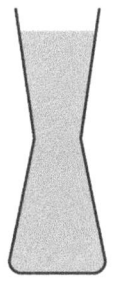

E.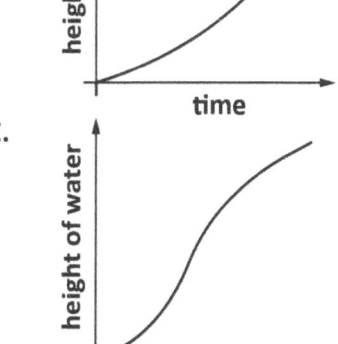

MAIN ACTIVITY: INTERPRETING GRAPHS   **TIMING: 15 MINS**

LEARNING OBJECTIVES

- Identify and interpret roots, intercepts, turning points and gradients of graphs, relating them to real-life problems

EQUIPMENT

- ruler

1. **This graph shows how a taxi company charges for journeys of up to 50 miles.**

   a) Discuss with your tutor what the gradient and *y*-intercept mean.

   gradient = ...

   ....

   y-intercept = ...

   ....

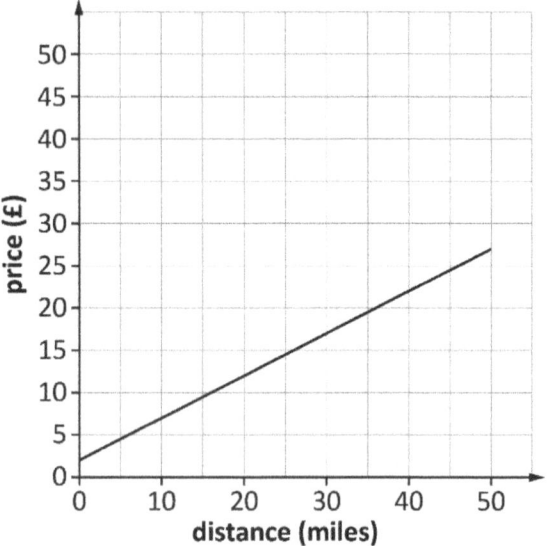

   b) A second company charges £1 per mile. There is no minimum fare.
      Draw a graph on the same axes to show the prices for this second company for journeys of up to 50 miles.

   c) Explain what the point of intersection of these two graphs represents.

2. **The relationship between the number of cars produced in a factory each week and the price of manufacture is shown in this graph.**

   a) Describe to your tutor what the shape of the graph tells you about the cost of production.

   ....

   ....

   b) How many cars should the company manufacture each week to ensure the lowest costs possible?

   ....

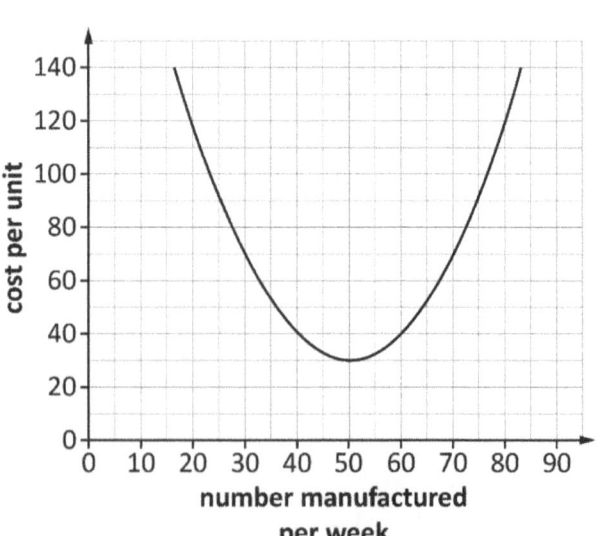

MAIN ACTIVITY: DISTANCE–TIME AND SPEED–TIME GRAPHS TIMING: 25 MINS

LEARNING OBJECTIVES

- Interpret graphs representing real-life situations

EQUIPMENT

- ruler

1. This distance–time graph shows Maddie's training run.

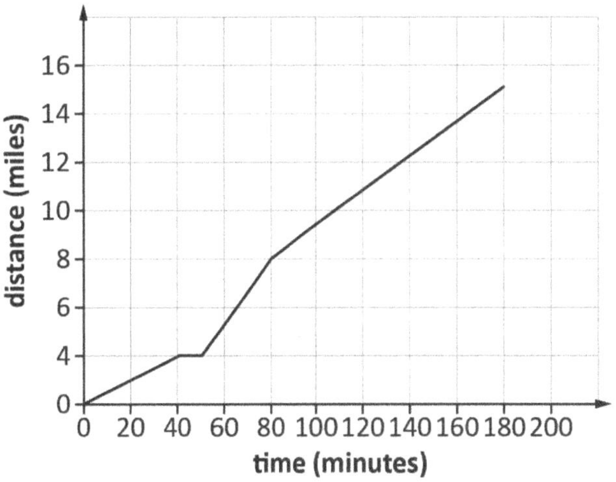

a) How far does she run in the first 40 minutes?

....

b) Work out her speed in miles per hour during the first 40 minutes of her run.

....

c) What happens between 40 and 50 minutes? Write down how you know.

....

d) When is Maddie running fastest? Give a reason for your answer and work out her speed.

....

e) How far did Maddie run altogether? ....

f) Work out Maddie's average speed for her run. ....

g) Simon does the same training run as Maddie. He starts at the same time as Maddie and runs at a constant speed of 6 mph. Show Simon's run on the graph.

h) After how many minutes does Maddie overtake Simon? ....

2. This graph shows the speed of a ball *t* seconds after it has been thrown upwards.

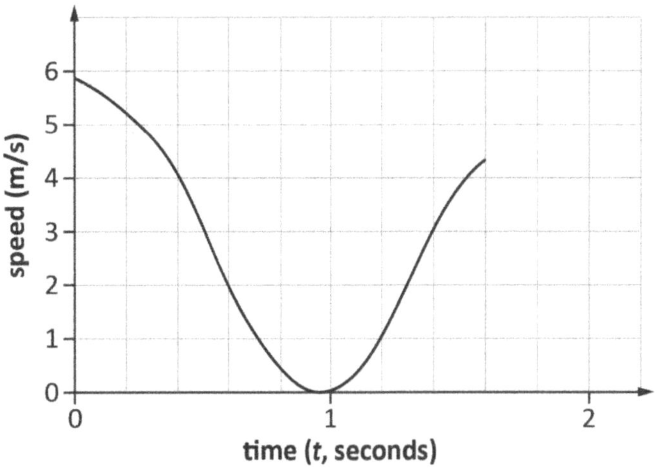

a) What is the speed of the ball 0.8 seconds after it has been thrown?

....

b) What was the lowest speed of the ball?

....

c) When was the ball travelling at 3 m/s?

....

HOMEWORK ACTIVITY: ALL TYPES OF GRAPH

TIMING: 60 MINS

LEARNING OBJECTIVES

- Interpret the reciprocal function
- Draw distance–time graphs and calculate the speed of individual sections
- Use the form $y = mx + c$ to identify parallel lines and y-intercept
- Find the equation of a line through two points or one point and gradient
- Find the equation of a straight line graph from the graph

EQUIPMENT

- graph paper
- ruler
- poster paper
- coloured pens or pencils
- revision cards

1. **Produce a set of revision cards or a poster including all the information you have learnt about linear graphs. You should include:**

   - how to give the gradient and y-intercept from the equation
   - how to work out the equation of a line from two points or one point and the gradient
   - how to identify when two lines are parallel from their equations
   - how to work out the equation of a line from the graph.

2. **Mr Murphy's mini triathlon consisted of a swim, cycle and run. This table shows the distances and the time it took him to complete each part of the triathlon.**
   Mr Murphy took a five minute break between each leg, and he travelled at a constant speed for each leg of the triathlon.

Leg	Time taken
1000 m swim	10 minutes
20 km cycle	1 hour 30 minutes
5 km run	20 minutes

   a) On graph paper, draw a distance–time graph to show his triathlon.

   b) How long did it take him to complete the mini triathlon? ....

   c) Work out his speed for each leg of the triathlon.

   swim: ....   cycle: ....   run: ....

3. **This is the graph of $y = \dfrac{1}{x}$.**

   Decide whether these statements are true or false.

   a) As x increases, y increases.   ....

   b) As x increases, y decreases.   ....

   c) The gradient of the graph is always negative.

   ....

   d) The gradient of the graph changes from negative to positive.

   ....

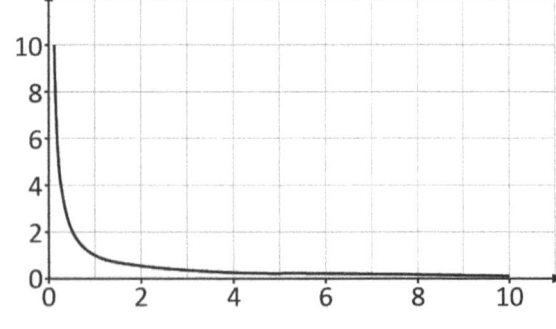

9 ANSWERS

STARTER ACTIVITY: FILLING VASES
1. B   2. C   3. A   4. D   5. E

MAIN ACTIVITY: INTERPRETING GRAPHS
1. a) *y*-intercept is the fixed cost (£2), gradient is the cost per mile (£0.50)

b) Check the student's graph. The line should pass through (0, 0) and (50, 50).

c) The distance for which both companies charge the same

2. a) The cost of production decreases for up to 50 cars, but then it starts to increase again.

b) 50

MAIN ACTIVITY: DISTANCE–TIME AND SPEED–TIME GRAPHS

1. a) 4 miles	b) 6 mph	c) She has a rest – the line is horizontal.	d) Between 50 and 80 minutes (8 mph)
e) 15 miles	f) 5 mph	g) Check the student's graph.	h) 68 minutes
2. a) 0.5 m/s	b) 0 m/s	c) 0.5 and 1.4 seconds	

HOMEWORK ACTIVITY: ALL TYPES OF GRAPH
1. Check the student's cards/poster.

2. a)

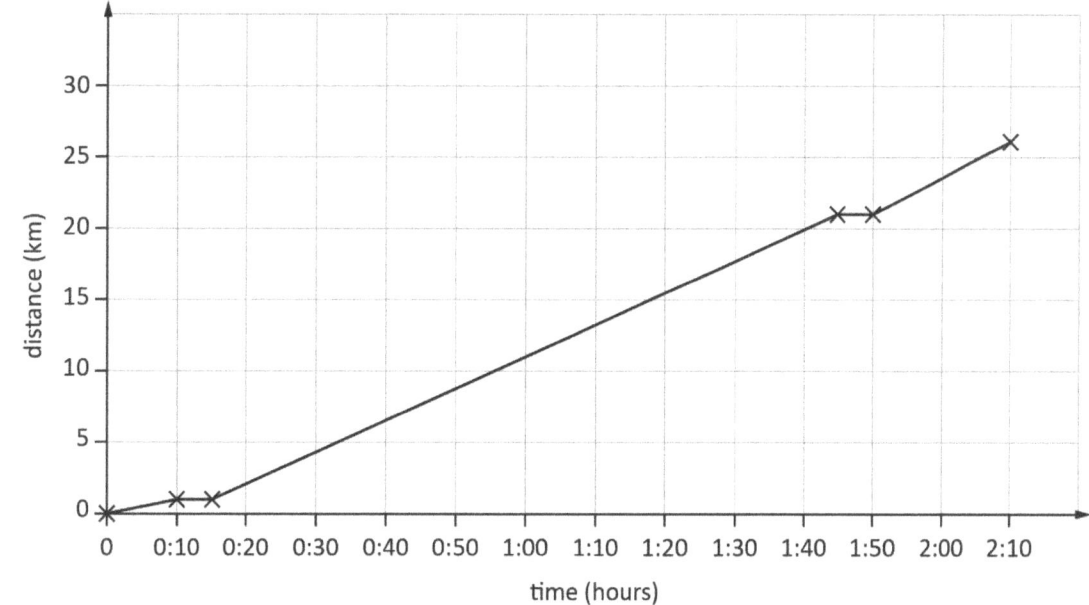

b) 2 hours 10 minutes

c) swim = 6 km/h   cycle = $13\frac{1}{3}$ km/h   run = 15 km/h

3. a) false   b) true   c) true   d) false

10 ALGEBRA: SOLVING LINEAR EQUATIONS

LEARNING OBJECTIVES

- Use and interpret algebraic notation
- Substitute numerical values into formulae and expressions
- Understand and use the concepts and vocabulary of expressions, equations, formulae, identity, terms
- Simplify and manipulate algebraic expressions
- Solve linear equations in one unknown algebraically

SPECIFICATION LINKS

- A1, A2, A3, A4, A17

STARTER ACTIVITY

- **Simplify the expression; 5 minutes; page 72**
  Check the student understands the term 'equivalent'; if necessary, confirm that every expression on the left simplifies to one of the expressions on the right.

MAIN ACTIVITIES

- **Simplify, substitute, solve; 30 minutes; page 73**
  Remind the student that algebraic terms follow the same conventions as numerical values. Work through questions 1 and 2. Discuss the meaning of the word 'equation', and establish the concept of ensuring equality by carrying out the same action to both sides of an equation. Reinforce that to solve an equation, we must first collect like terms on one side of the equation before carrying out inverse operations.
- **Words and expressions; 10 minutes; page 74**
  If necessary, begin with the words 'term', 'expression' and 'equation', using an example to illustrate each. Remind the student of the inequality and identity signs.

PLENARY ACTIVITY

- **Mind map; 5 minutes**
  Ask the student to draw a mind map of all the algebraic techniques they have covered in this lesson. Encourage them to identify the areas of algebra they find most challenging.

HOMEWORK ACTIVITY

- **How hot?; 30 minutes; page 75**
  All instructions are given on the sheet, but you may wish to adjust the challenge level to suit your student. For example, you could ask them to complete one row, one diagonal, or the whole sheet.

SUPPORT IDEA

- **Simplify, substitute, solve** When solving equations, draw out a pair of old fashioned scales, inviting the student to see the equals sign as indicating balance. Establish that to maintain this balance, the same must be added to or subtracted from both sides of the equation.

EXTENSION IDEA

- **Simplify, substitute, solve** Invite the student to design their own questions based on the format of either activity. You may wish to give them a specific answer to aim for.

PROGRESS AND OBSERVATIONS

STARTER ACTIVITY: SIMPLIFY THE EXPRESSION

TIMING: 5 MINS

LEARNING OBJECTIVES	EQUIPMENT
• Use and interpret algebraic notation	none

Draw lines to match each of the expressions on the left with its equivalent simplified expression on the right. Some of the expressions on the right will match more than one of the expressions on the left.

1. $a + a$

2. $4a^2 \div 2a$

3. $(2a)^2$

4. $\dfrac{4a}{a}$

5. $2a + 2a$

6. $4a \times a$

7. $a \times a$

8. $2 \times a^2$

9. $2a \times 2a$

10. $2a^2 \div a$

11. $4a \div 2$

12. $4a^2 \div 2a^2$

13. $\dfrac{4a^2}{2a^2}$

14. $2a \times a$

15. $2a \times 2$

2

a^2

$2a$

$2a^2$

$4a$

$4a^2$

4

MATHS
— FOUNDATION —

AQA

TUTORS' GUILD

MAIN ACTIVITY: SIMPLIFY, SUBSTITUTE, SOLVE **TIMING: 30 MINS**

LEARNING OBJECTIVES

- Simplify and manipulate algebraic expressions
- Substitute numerical values into formulae and expressions
- Solve linear equations in one unknown algebraically

EQUIPMENT

Worked example: Solve the equation: $\dfrac{3x+2}{4} = 2x - 7$

$$3x + 2 = 8x - 28$$
$$2 = 5x - 28$$
$$30 = 5x$$
$$6 = x$$

---

1. **Simplify each of these algebraic expressions.**

   a) $9y \times 3y$ ....

   b) $r^2 + r^2$ ....

   c) $5(2m - 5) + 4m$ ....

   d) $3mn - 12mn + 4m$ ....

   e) $2(x + 1) - 3(x - 5)$ ....

   f) $4y \times 5m \times 3a$ ....

2. **Substitute $a = 1$, $b = -3$ and $c = 11$ into each of these algebraic expressions.**

   a) $\dfrac{4c}{2}$ ....

   b) $3(2a + 1)$ ....

   c) $ab + b^2$ ....

   d) $8 - 2c$ ....

3. **The formula for area of a triangle is $A = \dfrac{1}{2}b \times h$. Work out the value of A when:**

   a) $b = 2$ cm, $h = 5$ cm ....

   b) $b = 0.1$ m, $h = 0.35$ m ....

4. **Solve the equations, showing all your working. Give your answers as fractions where necessary.**

   a) $3x + 1 = 5x - 7$ ....

   b) $3(x - 2) = -18$ ....

   c) $4x - 7 = x + 4$ ....

   d) $4x - 3 = 1 - 4x$ ....

   e) $2(3x + 1) = 10 - 5x$ ....

   f) $-3 = \dfrac{x + 4}{2}$ ....

MAIN ACTIVITY: WORDS AND EXPRESSIONS

TIMING: 10 MINS

LEARNING OBJECTIVES	EQUIPMENT
• Use and interpret algebraic notation	none
• Solve linear equations in one unknown algebraically	
• Understand and use the concepts and vocabulary of equations, formulae, identity, terms	

1. **Explain the meaning of each of these words to your tutor.**

   a) equation   b) identity   c) term   d) expression   f) inequality   g) formula

2. **Choose a word from the cloud below to describe each of the following.**

   a) $2x + 4$ ....   b) $4 < 2x + 5$ ....

   c) $4x \equiv x + x + x + x$ ....   d) $3a$ ....

   e) $y = mx + c$ ....   f) $2x - 5 = 17$ ....

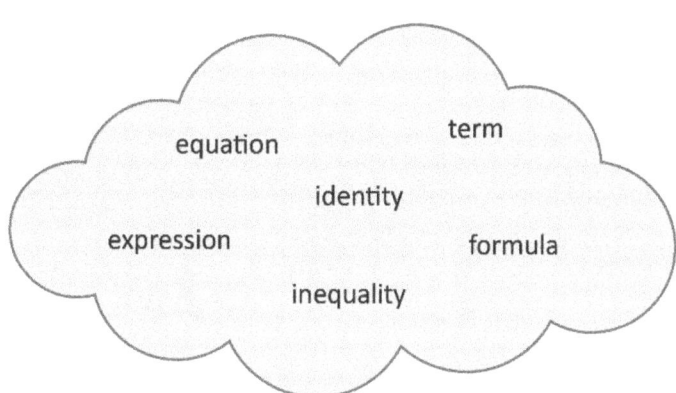

3. **Phoebe is two years older than her sister Alice. Alice is *b* years old.**

   a)   Write down an expression for Phoebe's age in terms of *b*.

   b)   The sum of Alice and Phoebe's ages is 42. Form and solve an equation to work out the ages of Alice and Phoebe.

HOMEWORK ACTIVITY: HOW HOT?

TIMING: 30 MINS

LEARNING OBJECTIVES

- Simplify and manipulate algebraic expressions
- Substitute numerical values into formulae and expressions
- Solve linear equations in one unknown algebraically

EQUIPMENT

1. **There are 16 challenges in the table below. Each row is a different level, each with a number of chillies – the more chillies, the 'hotter' the challenge.**

   You need to collect at least 16 chillies in total. Questions in the first row are worth one chilli, questions in the second row are worth two, and so on. Answers at least one question from each row.

Simplify	Substitute	Solve	Apply
$3x - 4x + 7x$	Work out the value of $4b - 2$ when $b = 5$.	$4x - 2 = 18$	My sister's age is 5 less than double mine. If I am x years old, write down an expression for my sister's age.
$3a \times 4a$	Work out the value of $3(2x - 5)$ when $x = 3$.	$x + 28 = 5(2x - 7)$	Form and solve an equation to calculate the value of x, and use this to calculate the size of the angles.
$3(4a - 1) + 6$	Work out the value of $x^2 - 3$ when $x = 7$.	$-2x + 4 = 10 - x$	Here is a number machine. If the input is a, the output is -9. Form and solve an equation to work out the value of a. $a \longrightarrow \boxed{\div 3} \longrightarrow \boxed{-7} \longrightarrow -9$
$3(4b - 2) - 2(b + 4)$	If $x = -2$ and $y = 4$, what is the value of $3x - 2y + 1$?	$\frac{1}{2}x + 12 = 2x + 3$	The formula for converting between degrees Celsius (C) and degrees Fahrenheit (F) is $F = \frac{9}{5}C + 32$. The temperature in Fahrenheit is 68°. Convert this into Celsius.

In the Apply / second row cell: angles labelled $x + 20°$, $x + 50°$, and $70°$.

10 ANSWERS

STARTER ACTIVITY: SIMPLIFY THE EXPRESSION

1. $2a$	2. $2a$	3. $4a^2$	4. 4	5. $4a$	6. $4a^2$	7. a^2
8. $2a^2$	9. $4a^2$	10. $2a$	11. $2a$	12. 2	13. 2	14. $2a^2$	15. $4a$

MAIN ACTIVITY: SIMPLIFY, SUBSTITUTE, SOLVE

1. a) $27y^2$ b) $2r^2$ c) $14m - 25$ d) $-9mn + 4m$ e) $-x + 13$ f) $60amy$

2. a) 22   b) 9   c) 6   d) -14

3. a) 5 cm^2 b) 0.0175 m^2

4. a) $x = 4$ b) $x = -4$ c) $x = \dfrac{11}{3}$ d) $x = \dfrac{1}{2}$ e) $x = \dfrac{8}{11}$ f) $x = -10$

MAIN ACTIVITY: WORDS AND EXPRESSIONS

1. See the glossary.

2. a) expression   b) inequality   c) identity   d) term   e) formula   f) equation

3. a) $b + 2$ b) $42 = 2b + 2$; Alice is 20 and Phoebe is 22.

HOMEWORK ACTIVITY: HOW HOT?

1.

Simplify	Substitute	Solve	Apply
$6x$	18	$x = 5$	$2x - 5$
$12a^2$	3	$x = 7$	$140 + 2x = 180$; 40° and 70°
$12a + 3$	46	$x = -6$	$a = -6$
$10b - 14$	-13	$x = 6$	$C = 20°$

GLOSSARY

Term
A single number or variable or product of a number and variable, e.g. y, $2x$

Expression
The sum of two or more algebraic terms, e.g. $3m + 7$, $19f - 2$

Equation
A statement that two mathematical statements are equal, e.g. $4s + 9 = 22$, $2g - 6 = 9$

Formula
A relationship between two or more variables, e.g. $y = mx + c$, $v = \dfrac{s}{t}$

Inequality
A statement that two mathematical statements are not equal, e.g. $4f < 6f$, $16 > 12 + c$

Identity
A statement that two mathematical statements are always equal, e.g. $2n + 3n = 5n$, $b \times b = b^2$

11 ALGEBRA: QUADRATIC EQUATIONS

LEARNING OBJECTIVES

- Use trial and improvement techniques to solve simple simultaneous equations
- Expand products of two binomials
- Factorise algebraic expressions by taking out common terms
- Factorise quadratic expressions including the difference of two squares
- Solve quadratic equations by factorising

SPECIFICATION LINKS

- A1, A4, A18

STARTER ACTIVITY

- **A and B?; 5 minutes; page 78**
  For each question, encourage the student to make an initial guess and go from there, writing down which numbers they have tried. Encourage them to notice that if the product of two numbers is zero, one must be zero.

MAIN ACTIVITIES

- **Expanding brackets; 20 minutes; page 79**
  Remind the student how to multiply one term by an expression in a bracket and extend to multiplying two sets of brackets.
- **Factorising and solving equations; 20 minutes; page 80**
  Encourage the student to identify the relationship between the positive and negative signs in the quadratic expression and its factorised form. Establish that Hamid's method works for all equations of the form $x^2 - ax + b$. If necessary, model solving a factorised quadratic expression. Explain that if the product is equal to zero, one of the brackets must be equal to 0.

PLENARY ACTIVITY

- **Solving equations graphically; 5 minutes**
  Ask the student to solve the equation $x^2 + x - 6 = 0$ by factorising. Then sketch the graph of $y = x^2 + x - 6$. Establish that when solving the equation, you are actually finding the roots of the quadratic (where it crosses the y-axis). Discuss how you might find an estimate of the solution to the equation $x^2 + 2x - 5 = 0$ (by plotting and identifying approximate roots). Sketch the graph to demonstrate this.

HOMEWORK ACTIVITY

- **Algebraic dominoes; 60 minutes; page 81**
  Full instructions are given on the activity sheet.

SUPPORT IDEA

- **Expanding brackets** You may wish to draw out a table to expand pairs of brackets, e.g. for $(x + 2)(x - 3)$:

	x	$+2$
x	x^2	$+2x$
-3	$-3x$	-6

EXTENSION IDEA

- **Factorising and solving equations** Ask the student to write a step-by-step method for factorising a quadratic expression. They may wish to use a flow chart to illustrate this.

PROGRESS AND OBSERVATIONS

STARTER ACTIVITY: A AND B?

TIMING: 5 MINS

LEARNING OBJECTIVES

- Use trial and improvement techniques to solve simple simultaneous equations

EQUIPMENT

1. If the sum of two numbers is 15 and the difference is 5, what are the two numbers?

....

2. $a \times b = 0$, but $a + b = 5$. What are a and b?

....

3. $\dfrac{a}{b} = 1$ and $a \times b = 25$. What are a and b?

....

4. If $a \times b = 0$, what can you say about either a or b?

....

MATHS
— FOUNDATION —

MAIN ACTIVITY: EXPANDING BRACKETS	TIMING: **20** MINS

LEARNING OBJECTIVES	EQUIPMENT
• Expand products of two binomials	none

 1. The area of this rectangle can be worked out in two ways.

method 1: $(4 + 3) \times (5 + 2)$
method 2: $(4 \times 5) + (4 \times 2) + (3 \times 5) + (3 \times 2)$

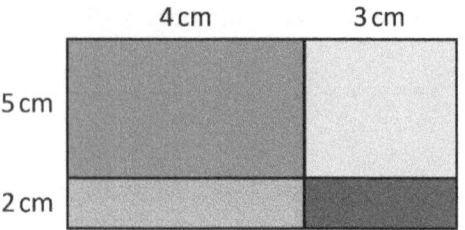

a) How do these methods work?

----

b) Check both methods give the same solution. --

We can say: $(4 + 3) \times (5 + 2) = (4 \times 5) + (4 \times 2) + (3 \times 5) + (3 \times 2)$.

 2. Using the method above, expand these pairs of brackets. Simplify your answers.

a) $(4 + 2)(3 + 7)$ ....

b) $(a + 4)(a + 2)$ ....

c) $(9 + 2)(7 - 3)$ ....

d) $(m - 1)(m + 5)$ ....

 3. Expand and simplify these pairs of brackets.

a) $(x + 2)(x + 3)$ ....

b) $(x + 1)(x - 1)$ ....

c) $(x - 1)(x + 4)$ ....

d) $(x - 3)(x + 3)$ ....

e) $(x - 4)(x - 3)$ ....

f) $(x + 5)^2$ ....

MAIN ACTIVITY: FACTORISING AND SOLVING EQUATIONS   TIMING: 20 MINS

LEARNING OBJECTIVES	EQUIPMENT
• Factorise algebraic expressions by taking out common terms	none
• Factorise quadratic expressions including the difference of two squares	
• Solve quadratic equations by factorising	

1. **Complete these factorisations.**

   a) $3x - 6y =$ .... $(x - 2y)$

   b) $2xy + 4x =$ .... $(y +$ ....$)$

   c) $4x + 12 =$ .... $(x +$ ....$)$

   d) $3m^2 - m = m ($..............$-$..............$)$

   e) $10 - 15y =$ .... $($ .... $-$ ....$)$

   f) $12xyz + 3xy =$ .... $($ .... $+$ ....$)$

2. **Draw lines to match each factorised expression to its expanded form.**

$(x + 2)(x + 3)$	$x^2 + 2x + 1$
$(x + 4)(x - 2)$	$x^2 - 7x + 6$
$(x - 1)(x + 1)$	$x^2 + 2x - 8$
$(x - 1)(x - 6)$	$x^2 - 1$
$(x + 1)^2$	$x^2 + 5x + 6$

3. **Hamid factorises $x^2 + 8x + 15$.**
   He thinks: *I need two numbers that multiply to give 15 and add to give 8.*

   a) Which two numbers satisfy this?   ....

   b) Complete this factorisation. $x^2 + 8x + 15 = (x +$ ....$)(x +$ ....$)$

   c) Use Hamid's method to factorise these expressions.

   i) $x^2 + 7x + 10 =$ ....

   ii) $x^2 + x - 12 =$ ....

   iii) $x^2 - 6x + 8 =$ ....

4. **Look at this expression: $x^2 + 5x + 6$**

   a) Factorise this expression.   ....

   b) Use your answer to part a) to solve the equation $x^2 + 5x + 6 = 0$.   ....

MATHS
— FOUNDATION —

AQA

TUTORS GUILD

HOMEWORK ACTIVITY: ALGEBRAIC DOMINOES   **TIMING: 60 MINS**

LEARNING OBJECTIVES
- Solve quadratic equations by factorising

EQUIPMENT
- scissors

1. Cut along the dotted lines to separate the dominoes. Make a chain by laying them end-to-end so that each quadratic expression is next to its factorised form. When you have finished, number the dominoes so you can show your tutor how you arranged them.

BEGIN	$(x + 1)(x - 1)$	$x^2 + x - 6$	$x^2 + 6x + 9$
$x^2 + 2x + 1$	$x^2 + 9x + 20$	$(x - 12)(x + 12)$	$x^2 + 2x - 3$
$x^2 - 7x + 10$	$(x - 5)(x + 2)$	$(x + 10)^2$	$(x - 9)(x - 3)$
$x^2 - 12x + 27$	$(x - 4)(x + 2)$	$x^2 - 1$	$(x + 3)(x - 2)$
$(x + 6)(x + 6)$	$x^2 - 144$	$x^2 - 2x - 8$	END
$(x - 5)(x + 2)$	$x^2 + 20x + 100$	$x^2 - 16$	$(x + 1)(x + 1)$
$(x + 3)(x - 1)$	$(x + 4)(x - 4)$	$(x + 3)^2$	$(x - 2)(x - 5)$
$x^2 - 3x - 10$	$x^2 + 12x + 36$	$(x + 4)(x + 5)$	$x^2 - 3x - 10$

2. Set each factorised quadratic expression on the dominoes equal to zero and solve for x.

....

....

....

11 ANSWERS

STARTER ACTIVITY: A AND B?

1. 10 and 5
2. 5 and 0
3. 5 and 5
4. Either a or b must be 0.

MAIN ACTIVITY: EXPANDING BRACKETS

1. a) method 1 – work out the total length and total width, then find the product

method 2 – work out the area of each of the smaller rectangles, then find the sum

b) $7 \times 7 = 49$; $20 + 8 + 15 + 6 = 49$

2. a) $(4 \times 3) + (4 \times 7) + (2 \times 3) + (2 \times 7) = 60$ b) $a^2 + 2a + 4a + 8 = a^2 + 6a + 8$

c) $(9 \times 7) + (9 \times -3) + (2 \times 7) + (2 \times -3) = 44$ d) $m^2 + 5m - m - 5 = m^2 + 4m - 5$

3. a) $x^2 + 5x + 6$ b) $x^2 - 1$ c) $x^2 + 3x - 4$ d) $x^2 - 9$ e) $x^2 - 7x + 12$ f) $x^2 + 10x + 25$

MAIN ACTIVITY: FACTORISING AND SOLVING EQUATIONS

1. a) $3x - 6y = 3(x - 2y)$ b) $2xy + 4x = 2x(y + 2)$ c) $4x + 12 = 4(x + 3)$

d) $3m^2 - m = m(3m - 1)$ e) $10 - 15y = 5(2 - 3y)$ f) $12xyz + 3xy = 3xy(4z + 1)$

2. $(x + 2)(x + 3) = x^2 + 5x + 6$ $(x + 4)(x - 2) = x^2 + 2x - 8$ $(x - 1)(x + 1) = x^2 - 1$

$(x - 1)(x - 6) = x^2 - 7x + 6$ $(x - 1)^2 = x^2 + 2x + 1$

3. a) 5 and 3   b) $x^2 + 8x + 15 = (x + 5)(x + 3)$

c) i) $(x + 5)(x + 2)$ ii) $(x + 4)(x - 3)$ iii) $(x - 4)(x - 2)$

4. a) $(x + 2)(x + 3)$ b) $x = -2$ or $x = -3$

HOMEWORK ACTIVITY: ALGEBRAIC DOMINOES

1.

BEGIN	$(x + 1)(x - 1)$	=	$x^2 - 1$	$(x + 3)(x - 2)$	=	$x^2 + x - 6$	$x^2 + 6x + 9$	=
$(x + 3)^2$	$(x - 2)(x - 5)$	=	$x^2 - 7x + 10$	$(x - 5)(x + 2)$	=	$x^2 - 3x - 10$	$x^2 + 12x + 36$	=
$(x + 6)(x + 6)$	$x^2 - 144$	=	$(x - 12)(x + 12)$	$x^2 + 2x - 3$	=	$(x + 3)(x - 1)$	$(x + 4)(x - 4)$	=
$x^2 - 16$	$(x + 1)(x + 1)$	=	$x^2 + 2x + 1$	$x^2 + 9x + 20$	=	$(x + 4)(x + 5)$	$x^2 - 3x - 10$	=
$(x - 5)(x + 2)$	$x^2 + 20x + 100$	=	$(x + 10)^2$	$(x - 9)(x - 3)$	=	$x^2 - 12x + 27$	$(x - 4)(x + 2)$	=
$x^2 - 2x - 8$	END							

2. $(x + 1)(x - 1) = 0$, $x = -1$ or $x = 1$ $(x + 6)(x + 6) = 0$, $x = -6$ $(x + 4)(x + 5) = 0$, $x = -4$ or $x = -5$

$(x + 3)(x - 2) = 0$, $x = -3$ or $x = 2$ $(x - 12)(x + 12) = 0$, $x = 12$ or $x = -12$ $(x - 5)(x + 2) = 0$, $x = 5$ or $x = -2$

$(x + 3)^2 = 0$, $x = -3$ $(x + 3)(x - 1) = 0$, $x = -3$ or $x = 1$ $(x + 10)^2 = 0$, $x = -10$

$(x - 2)(x - 5) = 0$, $x = 2$ or $x = 5$ $(x + 4)(x - 4) = 0$, $x = -4$ or $x = 4$ $(x - 9)(x - 3) = 0$, $x = 9$ or $x = 3$

$(x - 5)(x + 2) = 0$, $x = 5$ or $x = -2$ $(x + 1)(x + 1) = 0$, $x = -1$ $(x - 4)(x + 2) = 0$, $x = 4$ or $x = -2$

GLOSSARY

Factorise

The process of finding common factors

12 ALGEBRA: SIMULTANEOUS EQUATIONS

LEARNING OBJECTIVES

- Manipulate algebraic expressions
- Solve two linear simultaneous equations in two variables; find approximate solutions using a graph
- Translate simple situations into algebraic expressions; derive and solve two simultaneous equations and interpret the solution

SPECIFICATION LINKS

- A1, A2, A3, A4, A5, A9, A12, A19, A21

STARTER ACTIVITY

- **If I know… then what else do I know?; 5 minutes; page 84**
  Full instructions are given on the activity sheet.

MAIN ACTIVITIES

- **Simultaneous equations with algebra; 25 minutes; page 85**
  Question 1 guides the student through two methods for solving simultaneous equations. You may prefer to teach the student one method only; if so, model how to solve and move straight on to questions 2 and 3.
- **Simultaneous problems; 15 minutes; page 86**
  These questions require the student to apply the skills covered in the first part of the lesson. Encourage them to highlight the key information and to form new equations where necessary. In question 1, discuss how the graph relates to the equations $C = 5s + 50$ and $C = 2s + 120$, and the values of the variables at the point where the graphs intersect. Emphasise that trial and improvement techniques will not achieve any marks. Encourage the student to check their solutions by substituting into the original equations.

PLENARY ACTIVITY

- **Apples and pears; 5 minutes**
  Give the student the following problem and ask them to write an equation and solve it.
  Jill and Sanjay both buy some apples and some pears. Jill buys three apples and seven pears, and the cost is £3.91. Sanjay buys six apples and four pears and the cost is £4.12. How much do apples cost, and how much do pears cost? (Apples cost 44p and pears cost 37p.)

HOMEWORK ACTIVITY

- **Simultaneous equations; 60 minutes; page 87**
  Full instructions are given on the activity sheet.

SUPPORT IDEA

- **Simultaneous equations with algebra** Before asking the student to attempt questions 2 and 3, revise how to eliminate one unknown. Explain that if the signs are different, they add, and if the signs are the same, they subtract.

EXTENSION IDEA

- **Simultaneous equations with algebra** Invite the student to design their own pair of simultaneous equations to solve. Challenge them to make it one that you would wish to solve by elimination/substitution.

PROGRESS AND OBSERVATIONS

STARTER ACTIVITY: IF I KNOW... THEN WHAT ELSE DO I KNOW? TIMING: 5 MINS

LEARNING OBJECTIVES	EQUIPMENT
• Manipulate algebraic expressions	none

You are told that $a + b = 12$.

1. **Discuss with your tutor which of the following facts must be true. Make sure you can tell your tutor *why* you know they are true.**

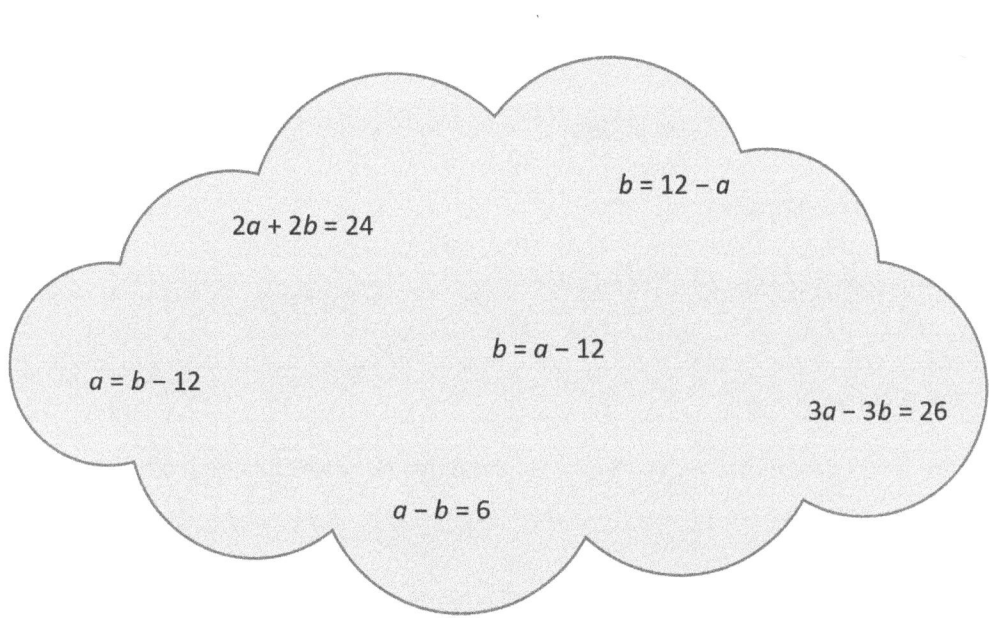

MAIN ACTIVITY: SIMULTANEOUS EQUATIONS WITH ALGEBRA TIMING: 25 MINS

LEARNING OBJECTIVES	EQUIPMENT
• Manipulate algebraic expressions	none

1. **Two students are asked to solve a pair of simultaneous equations:**

   $2x + y = 10$ $\qquad$ $4x - y = 8$

   Henry adds the two equations together and uses this to find x.

   $\begin{array}{r} 2x + y = 10 \\ +\ 4x - y = 8 \\ \hline 6x = 18 \end{array}$ $\qquad$ $x = 18 \div 6$
   $x = 3$

   a) Explain to your tutor why the y-term disappears when Henry adds together the two equations.

   b) If $x = 3$, work out the value of y. ....

   c) How could you check your values of x and y? ....

   d) Charlie decides to rearrange the first equation to make y the subject, and then substitute the expression for y into the second equation.

   $2x + y = 10$ $\qquad$ *$4x - (10 - 2x) = 8$* $\qquad$ Simplify and solve the equation to find the value of x.

   $y = 10 - 2x$ ....

   e) Substitute the value of x into Charlie's rearranged equation $y = 10 - 2x$ to find y. ....

   f) Which method do you prefer? Explain your reason to your tutor. ....

2. **Solve these simultaneous equations. Use your favourite method.**

   a) $x + y = 8$
   $\quad 2x + y = 6$

   b) $2x + y = 16$
   $\quad x + y = 9$

   c) $2x + 3y = 26$
   $\quad x - 3y = -32$

   $x =$ .... $\qquad$ $x =$ .... $\qquad$ $x =$ ....

   $y =$ .... $\qquad$ $y =$ .... $\qquad$ $y =$ ....

3. **Solve these simultaneous equations. You will need to multiply an equation before you can cancel the terms.**

   a) $2x + y = 9$
   $\quad x + 3y = 7$

   b) $x + y = 2$
   $\quad 3x - 2y = 16$

   c) $3x - y = 4$
   $\quad 2x + 2y = 0$

   $x =$ .... $\qquad$ $x =$ .... $\qquad$ $x =$ ....

   $y =$ .... $\qquad$ $y =$ .... $\qquad$ $y =$ ....

MAIN ACTIVITY: SIMULTANEOUS PROBLEMS

TIMING: 15 MINS

LEARNING OBJECTIVES

- Find approximate solutions to simultaneous equations using a graph
- Translate simple situations into algebraic expressions; derive and solve two simultaneous equations and interpret the solution

EQUIPMENT

- ruler

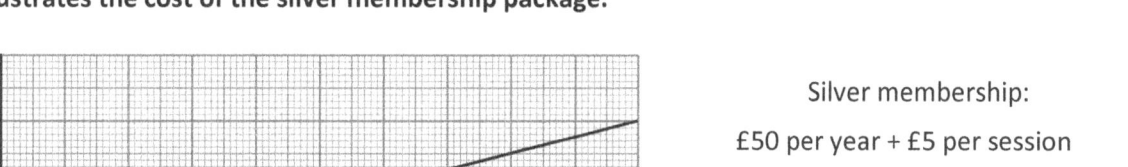

Example:

Alice buys two adult tickets and three children's tickets for a fairground ride. It costs her £29.50.

Wissam buys one adult ticket and one child's ticket for £12.50.

Work out the cost of one adult ticket.

$2A + 3C = 29.50$ (equation 1)

$A + C = 12.50$ (equation 2)

multiply equation 2 by 2: $2A + 2C = 25.00$ (equation 3)

equation 1 – equation 3: $C = 4.50$

substitute C into equation 2: $A + 4.50 = 12.50$ $A = 8$, so an adult ticket costs £8

---

1. **A gym offers two different membership packages, gold and silver.**
   This graph illustrates the cost of the silver membership package.

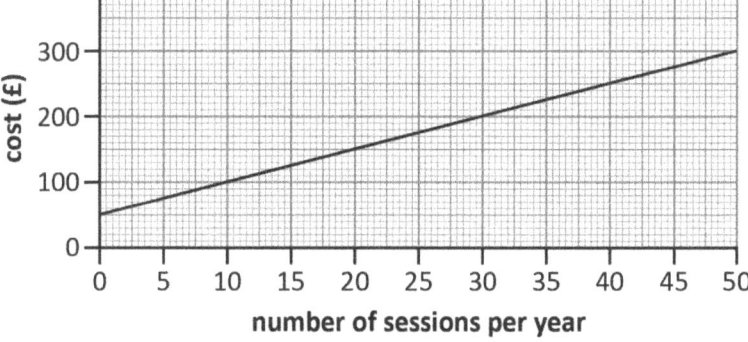

Silver membership:

£50 per year + £5 per session

Gold membership:

£120 per year + £2 per session

a) On the same axes, draw a graph to represent the cost of the gold membership.

b) Use the graph to work out the minimum number of sessions you'd have to attend in a year to make the Gold membership cheaper than the Silver membership.

---

2. **Two cups of tea and one coffee costs £6.50. Three cups of tea and two coffees cost £11.10.**
   How much would one cup of tea and one coffee cost?

MATHS
— FOUNDATION —

AQA

TUTORS' GUILD

HOMEWORK ACTIVITY: SIMULTANEOUS EQUATIONS

TIMING: 60 MINS

LEARNING OBJECTIVES

- Solve two linear simultaneous equations in two variables; find approximate solutions using a graph
- Translate simple situations into algebraic expressions; derive two simultaneous equations, solve the equations and interpret the solution

EQUIPMENT

- graph paper
- ruler

 1. In this activity the answers have been worked out for you – unfortunately they have been muddled up!
   Work out the value of *a* and *b* for each pair of simultaneous equations.
   Draw lines to match up the questions and answers.

questions

| $2a + b = 8$ | $a - b = -3$ | $3a + b = 3$ | $2a + b = 14$ | $a + b = -1$ |
| $3a - b = 2$ | $b - a = 3$ | $a + 3b = 17$ | $a - 2b = -3$ | $2a - 2b = -6$ |

answers

$a = -1$ $b = 3$ $a = 0$ $b = 1$ $a = -2$ $b = 4$ $a = 2$ $b = 6$ $a = 5$ $b = 4$

2. Solve these simultaneous equations by plotting the graphs. Check your answers algebraically.

a) $y = 2x, y = x - 5$ ....

b) $y = -2x + 1, y = 4x + 4$ ....

c) $y = 2x - 5, y = x$ ....

3. The menu in a coffee shop has been splashed and some of the prices are illegible.
   The waiter knows that two coffees and one hot chocolate cost £7.20, but three coffees and two hot chocolates cost £12.05. A tea costs £2.05. How much would one coffee, one tea and one hot chocolate cost?

   ....

   ....

12 ANSWERS

STARTER ACTIVITY: IF I KNOW... THEN WHAT ELSE DO I KNOW?

$2a + 2b = 24$ and $b = 12 - a$

MAIN ACTIVITY: SIMULTANEOUS EQUATIONS WITH ALGEBRA

1. a) $y + (-y) = 0$

b) $y = 4$

c) Substitute the values of x and y into both equations to check that they work.

d) $x = 3$

e) $y = 4$

f) Student's own answer

2. a) $x = -2, y = 10$ b) $x = 7, y = 2$ c) $x = -2, y = 10$

3. a) $x = 4, y = 1$ b) $x = 4, y = -2$ c) $x = 1, y = -1$

MAIN ACTIVITY: SIMULTANEOUS PROBLEMS

1. a) Check student's graph. Their line should pass through (0, 120) and (50, 220).

b) 24

2. £4.60

HOMEWORK ACTIVITY: SIMULTANEOUS EQUATIONS

1.

questions	$2a + b = 8$ $3a - b = 2$	$a - b = -3$ $b - a = 3$	$3a + b = 3$ $a + 3b = 17$	$2a + b = 14$ $a - 2b = -3$	$a + b = -1$ $2a - 2b = -6$
answers	$a = 2, b = 4$	$a = 0, b = 3$	$a = -1, b = 6$	$a = 5, b = 4$	$a = -2, b = 1$

2. Check student's graphs.

a) $x = -5, y = -10$ b) $x = -\dfrac{1}{2}, y = 2$ c) $x = 5, y = 5$

3. £6.90

GLOSSARY

Simultaneous equations

Equations involving two or more unknowns

13 ALGEBRA: INEQUALITIES

LEARNING OBJECTIVES

- Understand and use the concept and vocabulary of inequalities
- Solve linear inequalities in one variable; represent the solution set on a number line

SPECIFICATION LINKS

- A1, A2, A3, A4, A22

STARTER ACTIVITY

- **Always, sometimes, never; 5 minutes; page 90**
  Full instructions are given on the activity sheet.

MAIN ACTIVITIES

- **Representing inequalities; 20 minutes; page 91**
  Recap the meaning of the four inequality symbols, and discuss how they can be represented on a number line. Encourage the student to use the language 'closed' and 'open' when referring to unshaded and shaded circles at the end of the lines representing an inequality.

- **Solving inequalities; 20 minutes; page 92**
  Establish when an inequality stays true and when it doesn't, working through question 1. Discuss why multiplying or dividing both sides of an inequality by a negative number 'switches' the direction of the inequality sign.
  Model how to solve a linear inequality by working through the worked example. Make the link to solving linear equations.

PLENARY ACTIVITY

- **I think of a number; 5 minutes**
  Tell the student that you are thinking of a number. Four more than this number is less than 8. If the number is a positive whole number, what could it be? (3, 2, or 1). Ask the student to come up with their own question like this for you to try.

HOMEWORK ACTIVITY

- **Exam-style questions; 25 minutes; page 93**
  Full instructions are given on the activity sheet.

SUPPORT IDEAS

- **Solving inequalities** Start by modelling a more simple inequality such as $2x < 5$. Ask the student to suggest the values of x that would satisfy this, and identify the largest possible value of x. Ask the student to use this to represent the solution as an inequality.

EXTENSION IDEAS

- **Solving inequalities** Link the solution of an inequality to the graphical representation, e.g. for $2x + 5 < 3$, show the graphs of $y = 2x + 5$ and $y = 3$ and discuss how we could solve the inequality graphically.

PROGRESS AND OBSERVATIONS

STARTER ACTIVITY: ALWAYS, SOMETIMES, NEVER　　　TIMING: 5 MINS

LEARNING OBJECTIVES

- Understand and use the concept and vocabulary of inequalities

EQUIPMENT

1. Given that $a \leq b$, decide whether the following statements are always true, sometimes true, or never true.

Hint: If you are not sure, choose some possible values for a and b and test them. Don't forget that a and b could be positive, negative or zero.

$a + 2 \leq b + 2$　　...

$\dfrac{a}{2} < \dfrac{b}{2}$　　...

$a - 2 > b - 2$　　...

$-a < -b$　　...

$5a \leq 12b$　　...

$10 - a > 10 - b$　　...

$a = b$　　...

MAIN ACTIVITY: REPRESENTING INEQUALITIES

TIMING: 20 MINS

LEARNING OBJECTIVES

- Write down whole number values that satisfy an inequality
- Represent an inequality on a number line
- Construct an inequality to satisfy a set shown on a number line

EQUIPMENT

 1. **Choose numbers from the cloud which satisfy the inequalities shown:**

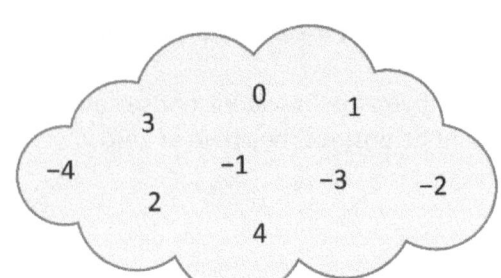

a) $x < -2$ ....

b) $-3 < x \leq 2$ ....

c) $x \geq -1.5$ ....

d) $-2.1 \leq x \leq -2$ ....

 2. **Inequalities can be represented on number lines.**

a) Explain to your tutor how you can represent an inequality on a number line. Make sure you explain when you would use 'closed' and 'open' circles.

b) Represent each of these inequalities on a number line.

i) $a < 2$

-5 -4 -3 -2 -1 0 1 2 3 4 5

ii) $-1 \leq x$

-5 -4 -3 -2 -1 0 1 2 3 4 5

iii) $-1.5 < b \leq 2.3$

-5 -4 -3 -2 -1 0 1 2 3 4 5

 3. **Write down the inequality represented on each of these number lines.**

a)

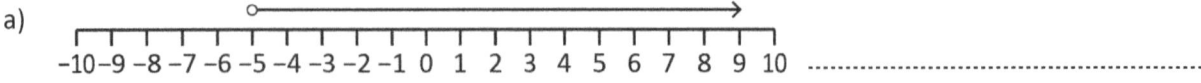

-10 -9 -8 -7 -6 -5 -4 -3 -2 -1 0 1 2 3 4 5 6 7 8 9 10 ....

b)

-10 -9 -8 -7 -6 -5 -4 -3 -2 -1 0 1 2 3 4 5 6 7 8 9 10 ....

c)

-10 -9 -8 -7 -6 -5 -4 -3 -2 -1 0 1 2 3 4 5 6 7 8 9 10 ....

d)

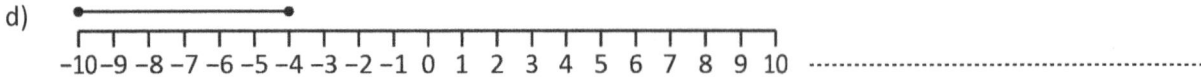

-10 -9 -8 -7 -6 -5 -4 -3 -2 -1 0 1 2 3 4 5 6 7 8 9 10 ....

MAIN ACTIVITY: SOLVING INEQUALITIES   TIMING: 20 MINS

LEARNING OBJECTIVES	EQUIPMENT
• Write down whole number values which satisfy an inequality • Solve linear inequalities in one variable; represent the solution set on a number line	none

> **Example:** Solve the inequality $2x + 1 \leq 7$.
> subtract 1 from both sides of the inequality to give $2x < 6$
> divide both sides by 2 to give $x < 3$
> $x \leq 3$

1. **Look at this inequality: 4 < 5. Answer the following questions and discuss your reasons with your tutor.**

   a) Will the inequality still be true if you:

      i) add any value to both sides? ....

      ii) subtract any value from both sides? ....

      iii) divide both sides by any positive number? ....

      iv) multiply both sides by any positive number? ....

   b) What happens to the inequality if you:

      i) multiply both sides by a negative number? ....

      ii) divide both sides by a negative number? ....

2. **Represent the solution to the example at the top of the page on the number line.**

   $-5 \quad -4 \quad -3 \quad -2 \quad -1 \quad 0 \quad 1 \quad 2 \quad 3 \quad 4 \quad 5$

3. **Solve each of these inequalities.**

   a) $3x + 4 > 25$ ....
   b) $10 \leq 2a - 5$ ....
   c) $0 < 5r - 3$ ....

4. **Answer these questions.**

   a) Solve the inequality $13 < 3x + 1$. ....

   b) Solve the inequality $3x + 1 < 34$. ....

   c) Discuss with your tutor how you could use your answers to a) and b) to write down the solution to the inequality $13 < 3x + 1 < 34$.

   d) Which integers satisfy the inequality $13 < 3x + 1 < 34$?

   ....

MATHS
— FOUNDATION —

AQA

TUTORS GUILD

HOMEWORK ACTIVITY: EXAM-STYLE QUESTIONS   **TIMING: 25 MINS**

LEARNING OBJECTIVES

- Use the correct notation to show inclusive and exclusive inequalities
- Write down whole number values that satisfy an inequality
- Solve linear inequalities in one variable; represent the solution set on a number line

EQUIPMENT

---

1. **Draw a line to match each inequality to its description.**

   x is between 1 and 5. $1 \leq x \leq 5$

   x is less than 5 and greater than or equal to 1. $1 \leq x < 5$

   x can take any value between 1 and 5, including 1 and 5. $1 \leq x \leq 5$

   x is less than 5 and the smallest value x can take is 1. $1 < x < 5$

   (2 marks)

2. **Write down all the positive integers which satisfy the inequality $a < 3$.**

   .... **(1 mark)**

3. **n is an integer. Given that $-1.2 \leq n < 3.5$, what is the:**

   a) smallest value it could be?   .... **(1 mark)**

   b) largest value it could be?   .... **(1 mark)**

4. **a) Solve the inequality $-5 < 2n - 4 \leq 8$.** .... **(3 marks)**

   b) Represent the solution set on the number line:

   ```
   ┌─┬─┬─┬─┬─┬─┬─┬─┬─┬─┬─┬─┬─┬─┬─┬─┬─┬─┬─┬─┬─┐
   -10 -9 -8 -7 -6 -5 -4 -3 -2 -1  0  1  2  3  4  5  6  7  8  9  10
   ```

   (2 marks)

5. **Look at this inequality: $2x - 5 \leq -11$. Decide whether these statements are true or false.**

   a) The largest value x could take is -3.   .... **(1 mark)**

   b) x must be smaller than -16.   .... **(1 mark)**

   c) x must be larger than -3.   .... **(1 mark)**

   d) The value of x must be between 0 and 3.   .... **(1 mark)**

13 Answers

Starter activity: Always, sometimes, never

$a + 2 \leq b + 2$: always true $\dfrac{a}{2} < \dfrac{b}{2}$: sometimes true $a - 2 > b - 2$: never true $-a < -b$: never true $5a \leq$

$12b$: always true $10 - a > 10 - b$: sometimes true $a = b$: sometimes

Main activity: Representing inequalities

1. a) −3, −4   b) −2, −1, 0, 1, 2   c) −1, 0, 1, 2, 3, 4   d) −2

2. a) Make sure student explains that ≥ and ≤ are represented by closed circles, and < and > are represented by open circles. Circles should be placed above the numbers in the number line with a line drawn to join them.

b) i) 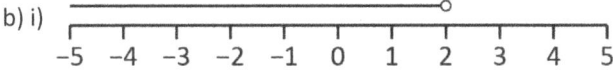   ii)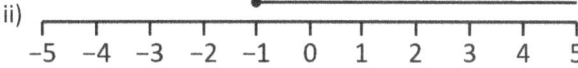

iii)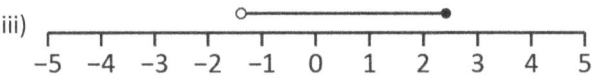

3. a) $x > -5$ b) $x \leq 1.5$ c) $-4 < x \leq 3$ d) $-10 \leq x \leq -4$

Main activity: Solving inequalities

1. a) i) yes   ii) yes   iii) yes   iv) yes
b) i) It is no longer true – its orientation switches.   ii) It is no longer true – its orientation switches.

2.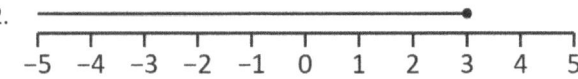

3. a) $x > 7$ b) $7.5 \leq a$ c) $\dfrac{3}{5} < r$

4. a) $4 < x$ b) $x < 11$ c) Combine the solutions: $4 < x < 11$ d) 5, 6, 7, 8, 9, 10

Homework activity: Exam-style questions

1. x is between 1 and 5: $1 < x < 5$
x is less than 5 and greater than or equal to 1: $1 \leq x < 5$
x can take any value between 1 and 5, including 1 and 5: $1 \leq x \leq 5$
x is less than 5 and the smallest value x can take is 1: $1 \leq x < 5$

2. 1, 2

3. a) −1   b) 3

4. a) $-\dfrac{1}{2} < n \leq 6$ b)

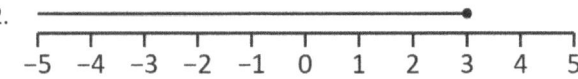

5. a) true   b) false   c) false   d) false

Glossary

Integer
A whole number. An integer can be positive, negative or zero.

14 ALGEBRA: SEQUENCES

LEARNING OBJECTIVES

- Generate terms of a sequence from a term-to-term or position-to-term rule
- Recognise and use sequences of triangular, square and cube numbers and simple arithmetic progressions
- Deduce expressions to calculate the nth term of a linear sequence

SPECIFICATION LINKS

- A1, A2, A23, A24, A25

STARTER ACTIVITY

- **Square and triangular numbers; 5 minutes; page 96**
  All instructions are given on the activity sheet. Ensure that to work out the 10th term, the student looks at the numerical pattern, i.e for triangular numbers: 1, 1 + 2, 1 + 2 + 3, ... or for square numbers: 1 × 1, 2 × 2, 3 × 3, ...

MAIN ACTIVITIES

- **Terms in a sequence; 20 minutes; page 97**
  When generating terms in a sequence, encourage students to look at the difference between terms to help them establish a pattern. Question 1d) is a Fibonacci style sequence. Discuss how these work with the student. Question 3 is the sequence of cube numbers, which students need to be familiar with.

- **Working out the nth term; 20 minutes; page 98**
  Explain that multiples of any number ($\pm x$) can be generated using the general term xn (i.e multiples of 4 can be generated using the general term $4n$). When looking at geometric sequences, ensure the student understands 'the ratio between consecutive terms' by demonstrating that each term divided by the previous term gives the same result.

PLENARY ACTIVITY

- **Types of sequence; 5 minutes**
  Brainstorm all the different types of sequence the student has encountered in the lesson. Ensure you include square, triangular, cube, Fibonacci, arithmetic and geometric. Ask the student to give an example of each type.

HOMEWORK ACTIVITY

- **Sequence problems; 30 minutes; page 99**
  Full instructions are given on the activity sheet.

SUPPORT IDEAS

- **Working out the nth term** For question 2, encourage the student to copy pattern 1 and then add the lines needed to make pattern 2, counting them carefully as they do so; repeat to make patterns 3 and 4. The general term is of the form $an + b$. Use the patterns to explain why the number of lines added each time is equal to a, then use the patterns to find b.

EXTENSION IDEAS

- **Working out the nth term** In question 3, challenge the student to find the general terms for the black tiles, the grey tiles and the total number of tiles.
- **Working out the nth term, Sequence problems** Ask the student to justify the general terms they find.

PROGRESS AND OBSERVATIONS

MATHS
— FOUNDATION —

AQA

Starter activity: Square and triangular numbers   Timing: 5 mins

Learning objectives	Equipment
• Recognise and use sequences of triangular and square numbers	none

1. Look at this pattern. It shows the triangular numbers.

a) Continue the sequence by drawing the next term above.

b) How many dots are there in each of the first five terms of the sequence?

....

c) Work out how many dots there will be in the 10th term.

....

2. Look at this pattern. It shows the square numbers.

a) Continue the sequence by drawing the next term.

b) How many dots are there in each of the first six terms of the sequence shown?

....

c) Work out how many dots there will be in the 10th term.

....

3. Why do you think these sequences are called the 'square numbers' and 'triangular numbers'?

....

....

MATHS

— FOUNDATION —

| MAIN ACTIVITY: TERMS IN A SEQUENCE | TIMING: 20 MINS |

LEARNING OBJECTIVES

- Generate terms of a sequence from a term-to-term or position-to-term rule
- Recognise and use sequences of triangular, square and cube numbers and simple arithmetic progressions

EQUIPMENT

The general term of a sequence is an algebraic expression used to calculate any term in the sequence.

Example:

The general term of a sequence is $2n - 5$. Give the first three terms of this sequence.

1st term: $2 \times 1 - 5 = -3$ 2nd term: $2 \times 2 - 5 = -1$ 3rd term: $2 \times 3 - 5 = 1$

---

1. **The first four terms of some sequences are given. For each sequence, give the next two terms and write them on the dotted lines. Explain to your tutor how you worked them out.**

   a) 2, 5, 8, 11, ..., ...   b) 27, 9, 3, 1, ..., ...

   c) 14, 10, 6, 2, ..., ...   d) 1, 1, 2, 3, ..., ...

2. **The first term of a sequence is −5. If the term-to-term rule is ×2 and then −2, what is the 4th term?**

   ....

3. **Look at this sequence:**   **1, 8, 27, 64, ...**

   a) Give the next term in the sequence.   ....

   b) How could you find any term in this sequence without finding all the terms in between?

   ....

4. **Give the first three terms and the 10th term of the sequences with these n^{th} terms:**

   a) $3n + 1$ 1st term: .... 2nd term: .... 3rd term: .... 10th term: ....

   b) $4n - 3$ 1st term: .... 2nd term: .... 3rd term: .... 10th term: ....

5. **This function machine generates terms in a sequence.**

   input ⟶ −5 ⟶ ×3 ⟶ output

   To work out the next term in the sequence, input the previous term. The first term of the sequence is 7.

   a) Give the 2nd term.   ....

   b) How many positive terms will there be in the sequence?   ....

MAIN ACTIVITY: WORKING OUT THE n^{TH} TERM   TIMING: **20** MINS

LEARNING OBJECTIVES

- Deduce expressions to calculate the nth term of a linear sequence

EQUIPMENT

For sequences where the the difference is constant, the general term will be in the form $an \pm b$.
You need to find a and b.

Example:
Work out the general term of the sequence: 3, 7, 11, 15, 19, …

Work out the difference between consecutive terms: $7 - 3 = 4$, $11 - 7 = 4$ … so the difference is +4. This is a.
Compare the sequence to the multiples of 4:   4, 8, 12, 16, 20, …
Each term in the sequence is 1 less than the corresponding term in the multiples of 4, so the general term is $4n - 1$.

---

1. **For each of these sequences, work out the nth term, and use it to give the 15th term of the sequence.**

   a) 12, 22, 32, 42, … n^{th} term = ...   15th term = ...

   b) 3, 1, −1, −3, … n^{th} term = ...   15th term = ...

2. **Eric makes a pattern using matchsticks.**
   Here are the first three patterns in a sequence:

   a)   Draw the next term in the sequence.

   b)   How many matchsticks will be needed for the 5th term in the sequence?   ....

   c)   Work out an algebraic expression for the number of matchsticks needed in the nth term of the pattern.

---

3. **A square pattern of tiles is made using grey and black tiles.**

   a)   How many black tiles will there be in the 4th pattern?   ....

   b)   Henry says *there cannot be 81 black tiles in a pattern*. Comment on whether Henry is correct.

An **arithmetic** sequence is a sequence in which the difference between consecutive terms is constant.
e.g. 10, 17, 24, 31, …
A **geometric** sequence is a sequence in which the ratio between consecutive terms is constant.
e.g 2, 4, 8, 16, 32, …

4. **Decide whether each of these sequences is arithmetic or geometric.**

   a) 10, 8, 6, …   ....   b) 10, 5, 2.5, …   ....

   c) −9, −4, 1 …   ....   d) 1, 3, 9, 27, …   ....

MATHS
— FOUNDATION —

HOMEWORK ACTIVITY: SEQUENCE PROBLEMS	TIMING: 30 MINS

LEARNING OBJECTIVES

- Generate terms of a sequence from a term-to-term or position-to-term rule
- Recognise and use sequences of triangular, square, cube numbers and simple arithmetic progressions
- Deduce expressions to calculate the nth term of a linear sequence

EQUIPMENT

1. **In a restaurant, one table seats four people. To seat more people, several tables can be put together as shown below.**

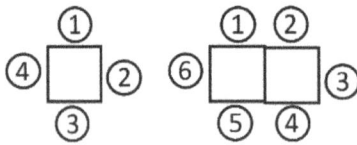

a) How many people would 3 tables seat? ....

b) Complete the table.

Number of tables	1	2	3	4	5	6
Number of seats	4					

c) Write down an expression for the number of seats there would be if n tables were put together.

---

d) How many tables would you need to seat 24 people? ....

2. **A sequence is generated by adding together three consecutive odd numbers.**

1st term = 1 + 3 + 5 = 9 2nd term = 3 + 5 + 7 = 15 3rd term = 5 + 7 + 9 = 21

a) What is the 12th term in the sequence? ....

b) Write down an expression for the nth term of the sequence.

---

c) Craig says that 95 cannot be a term in the sequence. Comment on whether Craig is correct.

14 Answers

Starter activity: Square and triangular numbers

1. a)    b) 1, 3, 6, 10, 15   c) 55 dots

2. a)    b) 1, 4, 9, 16, 25, 36   c) 100 dots

3. Student's own answers

Main activity: Terms in a sequence

1. a) 14, 17 (add 3)   b) $\frac{1}{3}$, $\frac{1}{9}$ (divide by 3)

c) −2, −6 (subtract 4)   d) 5, 8 (add together the last two terms)

2. −54

3. a) 125   b) These are the cube numbers, so the general term is n^3.

4. a) 4, 7, 10, 31   b) 1, 5, 9, 37

5. a) 6   b) 3

Main activity: Working out the n^{TH} term

1. a) $10n + 2$, 152   b) $-2n + 5$, −25

2. a)    b) 26   c) $5n + 1$

3. a) 128   b) Yes – the number of black tiles is always even.

4. a) arithmetic   b) geometric   c) arithmetic   d) geometric

Homework activity: Sequence problems

1. a) 8   b)

Number of tables	1	2	3	4	5	6
Number of seats	4	6	8	10	12	14

c) $2n + 2$ d) 11 tables

2. a) $23 + 25 + 27 = 75$ b) $6n + 3$

c) He is correct. If $6n + 3 = 95$, then $n = 15\frac{1}{3}$ which is not a whole number; or all the terms in the sequence are multiples of 3, and 95 is not a multiple of 3.

Glossary

Arithmetic sequence
A sequence in which the difference between consecutive terms is constant

Geometric sequence
A sequence in which the ratio between consecutive terms is constant

15 RATIO, PROPORTION AND RATES OF CHANGE: RATIO

LEARNING OBJECTIVES

- Use ratio notation and reduce a ratio to its simplest form
- Divide a quantity into a ratio
- Express a multiplicative relationship as a ratio or fraction
- Relate ratios to fractions or linear functions

SPECIFICATION LINKS

- R4, R5, R6, R7, R8, N11

STARTER ACTIVITY

- **Matching graphs; 5 minutes; page 102**
  Full instructions are given on the activity sheet.

MAIN ACTIVITIES

- **Bags of sweets; 20 minutes; page 103**
  Define what a ratio is and how it shows the relationship between two or more quantities. Full instructions are given on the activity sheet.
- **Working with ratios; 20 minutes; page 104**
  When cancelling ratios down, explain that all parts of the ratio must be in the same units before dividing through. Encourage your student to make links between ratio, fractions and proportion.

PLENARY ACTIVITY

- **Recipe ratios; 5 minutes**
  Write down a simple recipe (choose your own or find one online) and ask questions based around it.
  For example: *What is the ratio of flour to sugar? What fraction of the recipe is flour?*

HOMEWORK ACTIVITY

- **Practising ratio skills; 40 minutes; page 105**
  Full instructions are given on the activity sheet.

SUPPORT IDEAS

- **Working with ratios** Use physical objects to represent ratios. For a ratio of 2 : 3, collect five items and put them into two separate groups – one with two items and one with three. Show that you can sort multiples of five items in the same way: for every group of two, there must be a group of three.
- **Bags of sweets** Use a bar model to support the student's understanding of ratio. For example, if you have a ratio of 2 : 3, this can be represented pictorially. When sharing in a given ratio, each of the 'boxes' must be filled with the same amount. A bar model can be used for any ratio problems.

EXTENSION IDEAS

- **Bags of sweets** Ask the student to work out how many of each colour of sweet there are in each of these bags.
  There are 34 sweets in a bag. The bag contains green, red and orange sweets. The ratio of red to orange is 2 : 5, and the ratio of green to orange is 3 : 10. (8 red, 20 orange, 6 green)
  There are 25 sweets in a bag. The bag contains blue, pink and yellow sweets. There are twice as many pink sweets as blue sweets, and five more yellow sweets than blue sweets. (5 blue, 10 pink, 10 yellow)
  Challenge the student to write their own problem for you to solve. They must ensure they include enough information for the problem to have a unique answer.

PROGRESS AND OBSERVATIONS

MATHS
— FOUNDATION —

AQA

STARTER ACTIVITY: MATCHING GRAPHS

TIMING: 5 MINS

1. The statements below describe relationships between *x* and *y* values.
   Match each relationship with one of the graphs.

a) The *y* values are twice the *x*-values.

b) Each *x*-value is $\frac{1}{3}$ of the *y*-value.

c) *x* is three times larger than *y*.

d) The *y*-values are half the *x*-values.

e) The *x*-values are half the *y* values.

f) The *y*-values are three times the *x*-values.

g) The *y*-values are $\frac{1}{3}$ of the *x* values.

h) Each *x*-value is twice the *y*-value.

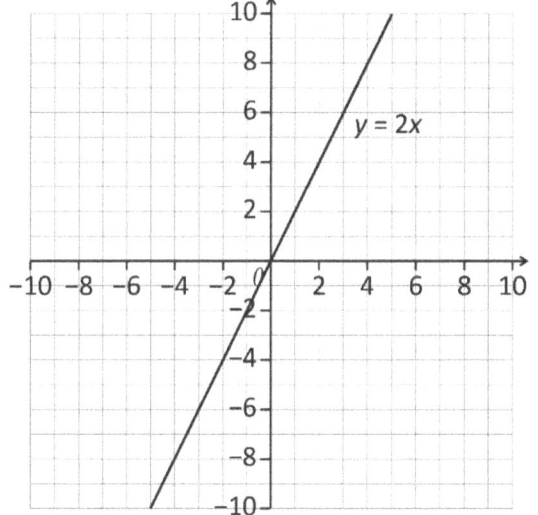

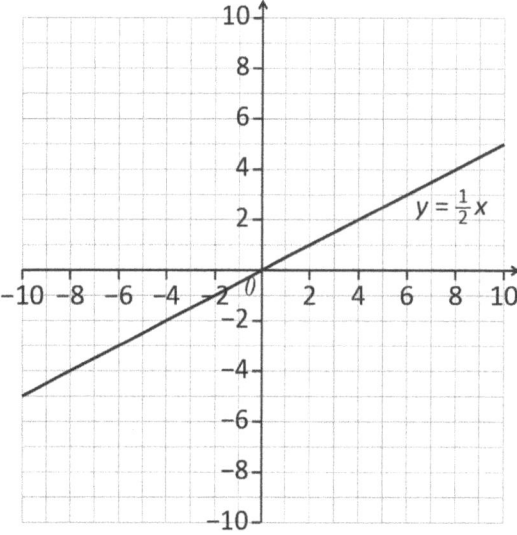

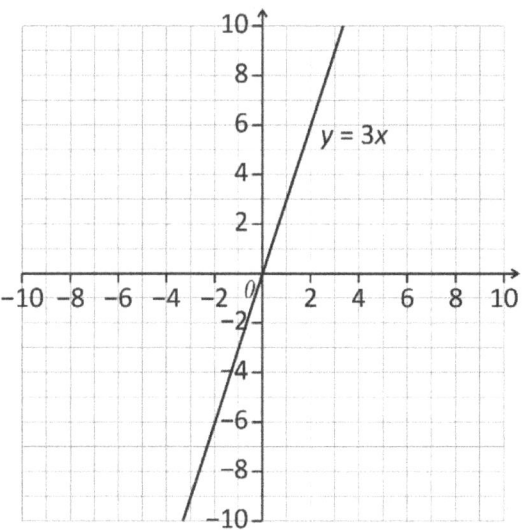

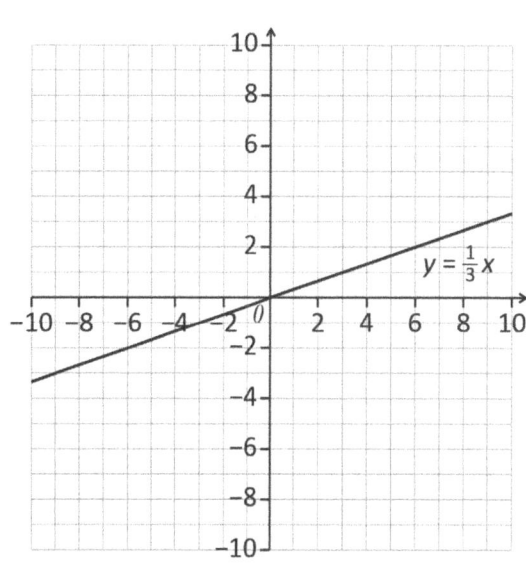

MAIN ACTIVITY: BAGS OF SWEETS

TIMING: **20** MINS

LEARNING OBJECTIVES

- Use ratio notation and reduce a ratio to its simplest form
- Divide a quantity into a ratio

EQUIPMENT

- bag of coloured sweets (such as jelly beans)

 1. **Open a bag of sweets and record how many of each colour sweet there are in the packet. Write five different sentences about the ratio of the colours of the sweets.**

For example, you might write 'the ratio of pink sweets to red sweets is 4 : 5'.

Make sure all your ratios are simplified as much as possible.

Colour	Frequency

----

2. **Some facts about bags of sweets are given in the table below.**
   Use these facts to complete the final column in the table.

Number of sweets	Clues	Number of sweets of each colour
24	There are only red and green sweets. The ratio of red to green sweets is 1 : 5.	red: green:
30	The bag contains red, yellow and green sweets. The ratio of red to yellow to green sweets is 1 : 2 : 3.	red: yellow: green:
15	The bag contains red, green and yellow sweets. The ratio of green to yellow is 1 : 2. There is 1 less red sweet than there are green sweets.	red: green: yellow:

MATHS
— FOUNDATION —

MAIN ACTIVITY: WORKING WITH RATIOS   **TIMING: 20 MINS**

LEARNING OBJECTIVES	EQUIPMENT

LEARNING OBJECTIVES
- Divide a quantity into a ratio
- Express a multiplicative relationship as a ratio or fraction
- Relate ratios to fractions or linear functions

EQUIPMENT
none

Example:

Paul and Clare share 20 tennis balls in the ratio 2 : 3. How many do they each get?

Work out how many 'parts' to share the tennis balls in.	2 + 3 = 5
Work out how many balls in each 'part'.	20 ÷ 5 = 4
Paul gets 2 parts:	2 × 4 = 8 balls
Clare gets 3 parts:	3 × 4 = 12 balls

---

1. **Share these amounts in the ratios given:**

   a) 450 ml in the ratio 4 : 5 _____   b) 2.5 m in the ratio 3 : 17 _____

Hint: A ratio can be simplified just like a fraction. As long as you divide all parts of the ratio by the same number the ratio is equivalent.

2. **Simplify these ratios.**

   a) 100 ml : 1 litre   _____

   b) 25 cm : 1.2 m   _____

   c) 2 hours : 1 hour 30 minutes   _____

3. **In a recipe, the weight of sugar to flour is in the ratio 2 : 3. If there is 50 g more flour than sugar, how much sugar is used?**

---

4. **In a school there are 250 students. 160 of them are boys. In class 1R, there are 22 boys and 15 girls. Is the proportion of boys and girls in the school the same as in class 1R? Explain your answer.**

---

5. **In a litter of puppies, two of the puppies are brown and the rest are white. The ratio of brown to white puppies is 1 : 3.**

   a) What fraction of the puppies are brown?   _____

   b) How many puppies are white?   _____

   c) Write down an equation which links the number of brown puppies (b) to the number of white puppies (w).

MATHS
— FOUNDATION —

AQA

HOMEWORK ACTIVITY: PRACTISING RATIO SKILLS

TIMING: 40 MINS

LEARNING OBJECTIVES

- Use ratio notation and reduce a ratio to its simplest form
- Divide a quantity into a ratio
- Express a multiplicative relationship as a ratio or fraction
- Relate ratios to fractions or linear functions

EQUIPMENT

- red, orange and green coloured pencils

The questions below each test you on a different skill involving ratios. Decide how you found each question and colour in the traffic light that applies to you.

red: I still find questions like this tricky   orange: I'm nearly there   green: I can do questions like this

 1. Using ratio notation and reducing a ratio to its simplest form

In a box of sweets, there are 7 red, 3 blue, 5 orange, 6 yellow, 4 green and 2 pink sweets. Write down the ratio of the following, giving each answer in its simplest form.

a) blue to red sweets   ....   b) green to red sweets   ....

c) pink to green sweets   ....   d) blue to yellow to red sweets   ....

e) orange to yellow sweets   ....   f) red to pink to green sweets   ....

 2. Sharing in a given ratio

   a) Amy and Rudi share £320 in the ratio 3 : 5. Work out how much they each receive.

---

   b) To make a salad dressing, Olivia mixes oil, honey and vinegar in the ratio 2 : 3 : 5. She makes 250 ml of salad dressing. How much honey does she use?

---

 3. Write a ratio as a fraction

   a) Angela and Sharon share a bag of sweets in the ratio 2 : 3. What fraction of the sweets does Sharon get?

---

   b) 200 ml of red paint, 350 ml of blue paint and 100 ml of white paint are mixed to make lilac paint.

     i)   Write down and simplify the ratio of red to blue to white paint.   ....

     ii) What fraction of the paint used is white?   ....

 4. Write a ratio as a linear function

The ratio of boys to girls in a school is 1 : 2. Write an equation which links the number of girls (g) to the number of boys (b).

---

15 ANSWERS

STARTER ACTIVITY: MATCHING GRAPHS

1. a) $y = 2x$ b) $y = 3x$ c) $y = \frac{1}{3}x$ d) $y = \frac{1}{2}x$

e) $y = 2x$ f) $y = 3x$ g) $y = \frac{1}{3}x$ h) $y = \frac{1}{2}x$

MAIN ACTIVITY: BAGS OF SWEETS

1. Student's own answers

2.

Number of sweets	Number of sweets of each colour
24	red: 4 green: 20
30	red: 5 yellow: 10 green: 15
15	red: 3 green: 4 yellow: 8

MAIN ACTIVITY: WORKING WITH RATIOS

1. a) 200 ml : 250 ml   b) 0.375 m : 2.125 m
2. a) 1 : 10   b) 5 : 24   c) 4 : 3
3. 100 g
4. No. The proportion of boys to girls in the school is 16 : 9, but the proportion of boys to girls in 1R is 22 : 15. Since the ratios are not equivalent, the proportions are not equal.
5. a) $\frac{1}{4}$ b) 6   c) $w = 3b$

HOMEWORK ACTIVITY: PRACTISING RATIO SKILLS

1. a) 3 : 7   b) 4 : 7   c) 1 : 2   d) 3 : 6 : 7   e) 5 : 6   f) 7 : 2 : 4
2. a) Amy = £120, Rudi = £200   b) 75 ml honey
3. a) $\frac{3}{5}$ b) i) 200 : 350 : 100 = 4 : 7 : 2   ii) $\frac{2}{13}$
4. a) $g = 2b$

16 RATIO, PROPORTION AND RATES OF CHANGE: DIRECT AND INVERSE PROPORTION

LEARNING OBJECTIVES

- Solve problems involving direct and inverse proportion including graphical and algebraic representations
- Understand that x is inversely proportional to y is equivalent to x is proportional to $\frac{1}{y}$

SPECIFICATION LINKS

- R1, R10, R11, R13

STARTER ACTIVITY

- **How long and how much?; 5 minutes; page 108**
  This activity introduces the ideas of inverse and direct proportion in a real-life scenario.

MAIN ACTIVITIES

- **Direct and inverse proportion; 15 minutes; page 109**
  Establish what direct and inverse proportion mean. Emphasise that direct proportion only occurs if when one value is zero, the other is also zero. Explain that, in inverse proportion, x is proportional to $\frac{1}{y}$.

- **Proportion in real life; 25 minutes; page 110**
  These problems are based on real-life scenarios. You may wish to encourage the student to use the unitary method to help solve them. Emphasise the need for common sense when solving problems involving proportion: *do you expect one thing to increase or decrease as the other increases or decreases?*

PLENARY ACTIVITY

- **Direct or inverse?; 5 minutes**
  Challenge the student to think of scenarios which involve direct and inverse proportion. Further challenge them to suggest an equation showing direct or inverse proportion and to sketch a graph illustrating them both.

HOMEWORK ACTIVITY

- **All aboard!; 45 minutes; page 111**
  Full instructions are given on the activity sheet.

SUPPORT IDEA

- **Proportion in real life** When working with proportion in real life, you may wish to ask the student to consider whether they should be multiplying or dividing. Encourage the student to think about whether their answer should be larger/smaller and which action will result in the desired solution.

EXTENSION IDEA

- **Proportion in real life** To extend the idea of proportionality, discuss with the student how you could convert between compound measures, such as litres per minute and millilitres per second.

PROGRESS AND OBSERVATIONS

 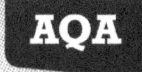
STARTER ACTIVITY: HOW LONG AND HOW MUCH?   TIMING: 5 MINS

LEARNING OBJECTIVES	EQUIPMENT
• Solve problems involving direct and inverse proportion	none

A wall is being built. A builder estimates it will take three men 12 hours to build it.

1. **How long will it take the following numbers of men to build the same wall?**

   a) one man   ....

   b) six men   ....

2. **How long will it take:**

   a) three men to build a wall twice the size?   ....

   b) two men to build a wall three times the size?   ....

   c) one man to build a wall four times the size?   ....

3. **If the builder charges £15 per man hour, how much will the original job cost if the following numbers of men build the wall?**

   a) three men   ....

   b) one man   ....

   c) eight men   ....

MATHS
— FOUNDATION —

MAIN ACTIVITY: DIRECT AND INVERSE PROPORTION   **TIMING: 15 MINS**

LEARNING OBJECTIVES

- Recognise direct and inverse proportion

EQUIPMENT

1. **Decide whether the following scenarios, graphs and equations show direct or inverse proportion. Fill in the gap with either direct or inverse. You must be able to justify your answers to your tutor.**

Remember: **Direct proportion** – as one value increases, the other increases; when one is zero, the other is zero.
Inverse proportion – as one value increases, the other decreases.

a) How much you earn is in .... proportion to the number of hours you work.

b) How long a job takes is in .... proportion to the number of people working on it.

c) The time it takes to run a mile is in .... proportion to the speed you are running.

d) The time it takes to answer maths questions is in .... proportion to the number of questions there are.

e)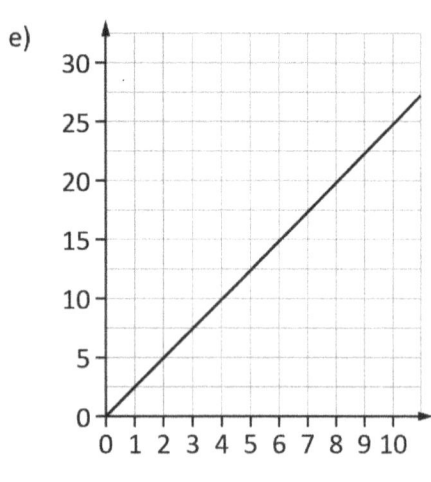

.... proportion

f)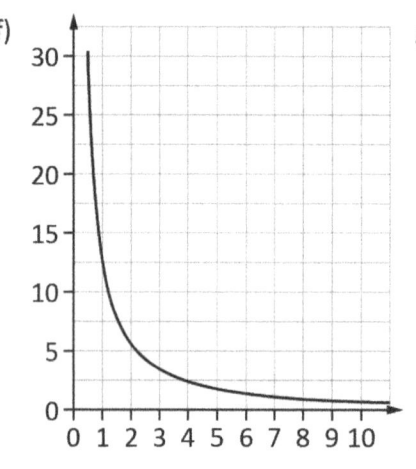

.... proportion

g)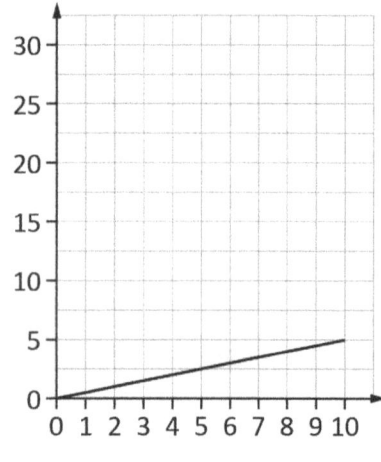

.... proportion

h) $y = 3x$ .... proportion

i) $y = 10x$ .... proportion

j) $y = \dfrac{0.3}{x}$ .... proportion

k) $y = 0.01x$ .... proportion

l) As x increases, y increases.   .... proportion

m) An increase in one value causes an increase in another value.   .... proportion

n) An increase in one value causes a decrease in another value.   .... proportion

o) As x increases, y decreases.   .... proportion

MAIN ACTIVITY: PROPORTION IN REAL LIFE

TIMING: 25 MINS

LEARNING OBJECTIVES	EQUIPMENT

- Solve problems involving direct and inverse proportion including graphical and algebraic representations

1. **Frank is going on holiday to Abu Dhabi. The exchange rate is £1 = 22 AED.**

   a) Work out how many AED he would get for £350. ....

   b) When he returns home, he has 148 AED left. He exchanges this into pounds. How much does he have left in pounds?

   ....

   c) A bottle of aftershave costs £27.75 in the UK and 445 AED in Abu Dhabi. Where is it cheaper?

   ....

2. **The ingredients for 20 meringues are 3 eggs and 180 g of sugar.**

   a) How much of each ingredient would be needed for 50 meringues? ....

   b) Tick the equation that accurately represents the relationship between the number of eggs (e) and the weight of sugar (s)?

   $s = 60e$ $\qquad$ $e = 60s$ $\qquad$ $e = \dfrac{60}{s}$ $\qquad$ $s = \dfrac{60}{e}$

3. **A tap runs at a rate of 12 litres per minute. 1 gallon ≈ 4.62 litres**

   a) How long will it take to run a bath containing 100 litres? ....

   b) Work out the rate of flow in litres per hour. ....

   c) Work out the rate of flow in gallons per hour. Give your answer to 3 significant figures.

   ....

4. **This graph shows water cooling over time.**
   Use the graph to work out:

   a) the temperature of the water 10 minutes after it was boiled

   ----

   b) the relationship between the temperature of the water and time after it has boiled.

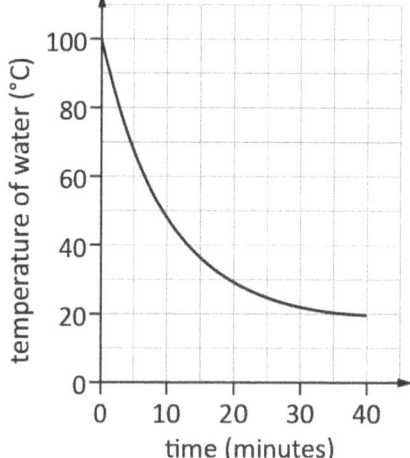

MATHS
— FOUNDATION —

HOMEWORK ACTIVITY: ALL ABOARD!   **TIMING: 45 MINS**

LEARNING OBJECTIVES

- Solve problems involving direct and inverse proportion including graphical and algebraic representations
- Understand that x is inversely proportional to y is equivalent to x is proportional to $\frac{1}{y}$

EQUIPMENT

- graph paper
- ruler

1. **The graph below shows the distance travelled by a train over a three hour journey.**

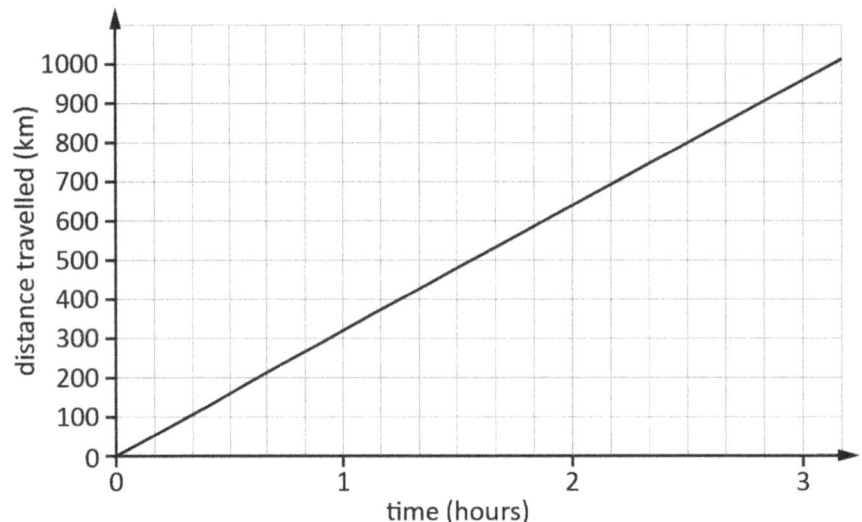

a) How far has the train travelled after 1.5 hours?   ....

b) Work out the speed of the train in km per hour.   ....

c) How long would the train take to travel 2000 km?   ....

d) How far did the train travel in 15 minutes?   ....

2. **The train will make a journey of 500 km without stopping.**

a) Complete the table of values showing the length of time taken for different speeds:

Speed (km/hour)	50	100	150	200	250	300	350
Time taken (hours)							

b) On graph paper, plot the graph of speed against time.

c) Use the graph to estimate how long it would take the train to travel the 500 km at a speed of 120 km/h.

d) Describe the relationship between the speed of the train and the time taken.

16 ANSWERS

STARTER ACTIVITY: HOW LONG AND HOW MUCH?

1. a) 36 hours   b) 6 hours
2. a) 24 hours   b) 54 hours   c) 144 hours
3. a) £540   b) £540   c) £540

MAIN ACTIVITY: DIRECT AND INVERSE PROPORTION

1. a) direct	b) inverse	c) inverse	d) direct	e) direct	f) inverse
g) direct	h) direct	i) direct	j) inverse	k) inverse	l) direct
m) direct	n) inverse	o) inverse			

MAIN ACTIVITY: PROPORTION IN REAL LIFE

1. a) 7700 AED   b) £6.73   c) Abu Dhabi
2. a) 7.5 eggs and 450 g sugar   b) $s = 60e$
3. a) 8 minutes 20 seconds   b) 720 litres per hour   c) 156 gallons per hour
4. a) 48 °C
b) As the time after boiling increases, the temperature of the water decreases. The two things are in inverse proportion.

HOMEWORK ACTIVITY: ALL ABOARD!

1. a) 480 km   b) 320 km/h   c) 6.25 hours   d) 80 km

2. a)

Speed (km/hour)	50	100	150	200	250	300	350
Time taken (hours)	10 hours	5 hours	$3\frac{1}{3}$ hours	$2\frac{1}{2}$ hours	2 hours	$1\frac{2}{3}$ hours	$1\frac{3}{7}$ hours

b)

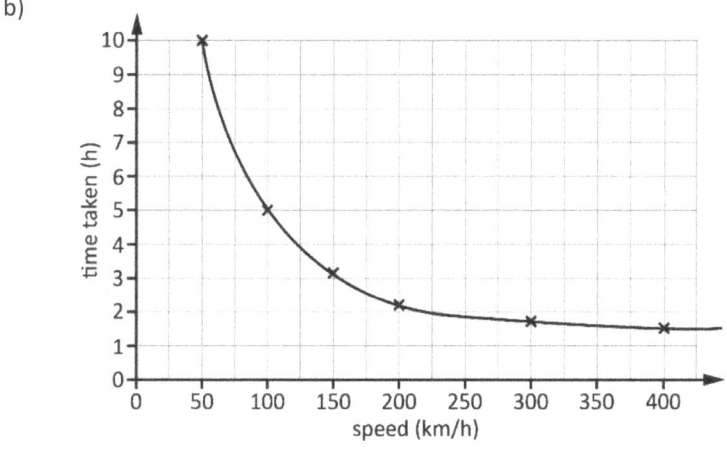

c) 4 hours 10 minutes
d) As the speed of the train increases, the time taken for the journey decreases. They are in inverse proportion.

GLOSSARY

Direct proportion

Two quantities are in direct proportion when both quantities increase at the same rate. An equation relating the two is of the form $y = kx$ where k is constant.

Inverse proportion

Two quantities are in inverse proportion when one quantity increases at the same rate as the other quantity decreases.

An equation relating the two is of the form $y = \frac{k}{x}$, where k is constant.

17 RATIO, PROPORTION AND RATES OF CHANGE: PROPORTION GRAPHS

LEARNING OBJECTIVES

- Interpret the gradient of a straight line graph as rate of change; recognise and interpret graphs that illustrate direct and inverse proportion
- Set up, solve and interpret the answers in growth and decay problems including compound interest

SPECIFICATION LINKS

- R14, R16, A10, A14

STARTER ACTIVITY

- **Working out percentages; 5 minutes; page 114**
  This activity should be completed without a calculator.

MAIN ACTIVITIES

- **What does the gradient represent?; 15 minutes; page 115**
  Recap that the gradient represents rate of change, and discuss that this means different things depending on what the graph represents. You could link back to Lesson 9 where you considered different types of graph representing real-life scenarios.
- **Growth and decay; 25 minutes; page 116**
  Remind the student of the idea of growth and decay, and explain the difference between compound and simple interest, linking back to lesson 4. Ask them to complete the questions.

PLENARY ACTIVITY

- **Sketching graphs; 5 minutes**
  Ask the student to sketch two graphs, one showing growth and one showing decay.

HOMEWORK ACTIVITY

- **Exam-style questions; 30 minutes; page 117**
  Full instructions are given on the activity sheet.

SUPPORT IDEAS

- **What does the gradient represent?** When considering the gradient of the graph, remind the student that they need to calculate the change in the 'y' value when the 'x' value changes by one. Encourage them to think logically about what is changing over time, and to express their answer in 'real-life' terms.
- **Growth and decay** Encourage the student to draw up a table of values when calculating compound interest.

EXTENSION IDEA

- **Growth and decay** Challenge the student to come up with a way of calculating compound interest over any number of years (e.g. 10% compound interest over n years with an initial deposit of A is $A \times 1.1^n$).

PROGRESS AND OBSERVATIONS

STARTER ACTIVITY: WORKING OUT PERCENTAGES

TIMING: 5 MINS

LEARNING OBJECTIVES

- Find percentages using mental methods

EQUIPMENT

1. **Each of these items has 40% off the original price. Calculate the sale prices.**

   a)   a shirt that originally cost £75

   ....

   b)   a pair of boots that originally cost £120

   ....

   c)   a pair of jeans that originally cost £85

   ....

2. **If the price of a coat in the sale is £120, what was the original price?**

MATHS
— FOUNDATION —

MAIN ACTIVITY: WHAT DOES THE GRADIENT REPRESENT?   TIMING: 15 MINS

LEARNING OBJECTIVES

- Interpret the gradient of a straight line graph as rate of change; recognise and interpret graphs that illustrate direct and inverse proportion

EQUIPMENT

1. **This graph shows the distance a train travels during a five hour journey.**

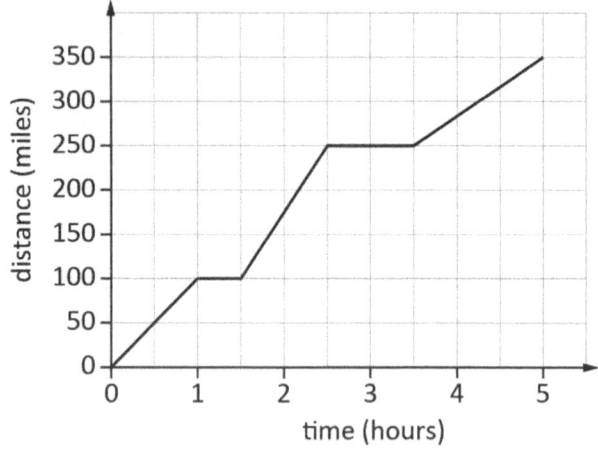

a) Complete this statement:

On a distance–time graph, the _____ can be calculated by finding the _____ .

b) When is the train travelling fastest? Give a reason for your answer.

....

c) Work out the speed for each part of the journey.

....

2. **For each scenario, find the gradient of the graph and explain what the gradient represents.**

a) depth of water in a bath after *t* seconds

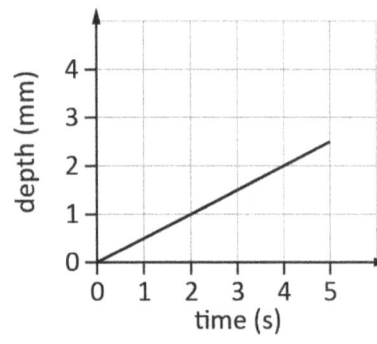

b) price of a taxi journey based on distance

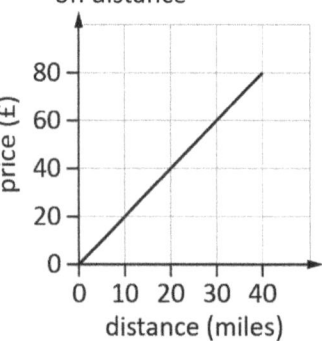

c) speed of a braking car over time

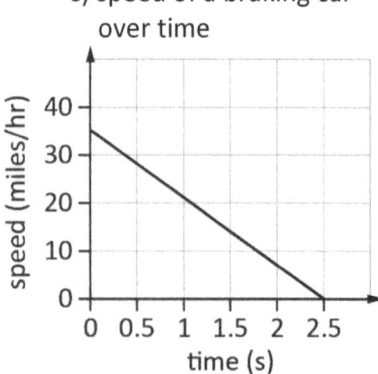

3. **Do any of the graphs in question 2 show direct proportion? If so, which ones?**

....

MAIN ACTIVITY: GROWTH AND DECAY   TIMING: 25 MINS

LEARNING OBJECTIVES

- Set up, solve and interpret the answers in growth and decay problems including compound interest

EQUIPMENT

- graph paper
- ruler
- calculator

1. **A bank account offers 5% compound interest. Erin invests £1000 in the bank.**

   a) How much will she have in the bank after:

   i) 1 year ....   ii) 2 years ....   iii) 3 years ....

   b) Complete this table.

Years invested	0	1	2	3	4	5	6
Total in account	£1000						

   c) On a piece of graph paper, draw a suitable pair of axes and plot a graph to show the amount Erin will have in her bank over a six-year period.

   d) How long will it be before Erin has over £1,500 in the bank? ....

2. **A bank has two different accounts.**

Account A	Account B
4% simple interest on all deposits	3% compound interest on all deposits

   Andrew has £500 to invest for 10 years. Which account should he invest it in? Explain your reason and show calculations to support it.

   ----

3. **The number of bacteria in a petri dish halves each half hour. Initially, there are 2.4×10^5 bacteria in the dish. How many will there be after three hours?**

MATHS
— FOUNDATION —

HOMEWORK ACTIVITY: EXAM-STYLE QUESTIONS	TIMING: **30** MINS

LEARNING OBJECTIVES

- Interpret the gradient of a straight line graph as rate of change; recognise and interpret graphs that illustrate direct and inverse proportion
- Set up, solve and interpret the answers in growth and decay problems including compound interest

EQUIPMENT

- calculator

1. **This graph shows the volume of water in a bath plotted against time.**

   a) Work out the gradient of the graph.

   .... **(2 marks)**

   b) What does the gradient represent?

   .... **(1 mark)**

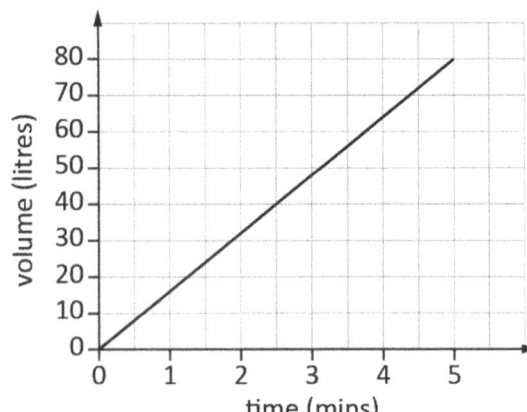

2. **The number of hits on a website increases over time.**
   Erica estimates that each day the number of hits increases by 10%.
   On the first day, the website had 2500 hits. How many hits would it have on the fourth day?
   Give your answer to the nearest whole number.

   .... **(3 marks)**

3. **Mr Sage spends £3000 on a motorbike. Its value depreciates at a rate of 3% per year.**
   How much will it be worth in three years' time? Give your answer to the nearest pound.

   .... **(3 marks)**

4. **The table shows the value of a house over a four-year period.**

Year	2010	2011	2012	2013
Value	£100 000	£110 000	£121 000	£133 100

   a) Show that the value of the house increases by the same proportion each year.

   .... **(2 marks)**

   b) Work out the percentage increase in value each year.

   .... **(2 marks)**

17 Answers

STARTER ACTIVITY: WORKING OUT PERCENTAGES

1. a) £45   b) £72   c) £51
2. £200

MAIN ACTIVITY: WHAT DOES THE GRADIENT REPRESENT?

1. a) On a distance–time graph, the **speed** can be calculated by finding the **gradient**.
b) Between 1.5 and 2.5 hours; the gradient is steepest.   c) 100 mph, 0 mph, 150 mph, 0 mph, 66.67 mph
2. a) Gradient = 0.5. Gradient represents rate of flow of water into the bath – the depth of water increases by 0.5 mm/s.
b) Gradient = 2. Gradient represents per mile or rate of increase in price – £2 per mile.
c) Gradient = 14. Gradient represents deceleration or rate of change of speed – the car decelerates at a rate of 14 mph/s.
3. Graphs a) and b)

MAIN ACTIVITY: GROWTH AND DECAY

1. a) i) £1050   ii) £1102.50   iii) £1157.63

b)

Years invested	0	1	2	3	4	5	6
Total in account	£1000	£1050	£1102.50	£1157.63	£1215.51	£1276.28	£1340.10

c)

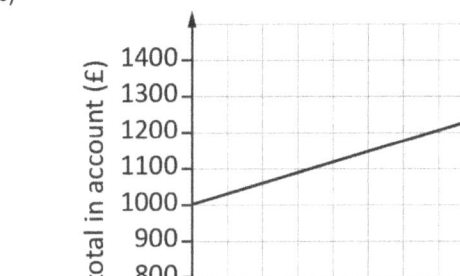
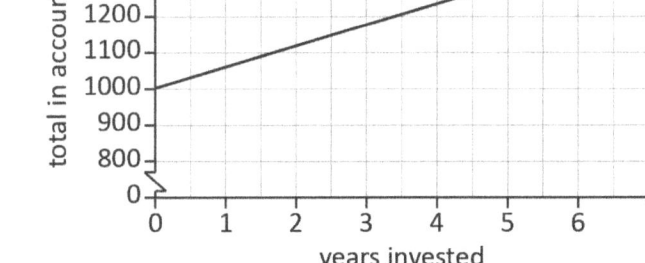

d) 9 years
2. Account A: 500 + 10(0.04 × 500) = £700; Account B: 500 × 1.03^{10} = £671.96 so Andrew should invest his money in Account A.
3. 3750

HOMEWORK ACTIVITY: EXAM-STYLE QUESTIONS

1. a) 16   b) The rate of flow of water into the bath (16 litres per minute)
2. 3328
3. £2,738
4. a) 110 000 ÷ 100 000 = 121 000 ÷ 110 000 = 133 100 ÷ 121 000 = 1.1 (or equivalent answer)
b) 10%

GLOSSARY

Simple interest
Interest calculated on the original investment only, not on any interest earned

Compound interest
Interest calculated on the original investment and any interest earned previously

18 GEOMETRY AND MEASURES: ANGLES

LEARNING OBJECTIVES

- Apply the properties of angles at a point, on a line, vertically opposite, corresponding and alternate
- Calculate and use the sum of angles in polygons

SPECIFICATION LINKS

- G1, G3

STARTER ACTIVITY

- **Colour it in; 5 minutes; page 120**
  Full instructions are given on the activity sheet.

MAIN ACTIVITIES

- **Using angle properties; 15 minutes; page 121**
  Remind the student of the properties of angles on a straight line, round a point and in a triangle, vertically opposite angles, corresponding and alternate angles, and co-interior angles. Ask the student to work through the activity.
- **Polygons; 25 minutes; page 122**
  For question 1, encourage the student to calculate the sum of the internal angles of the polygons by splitting them into triangles. Show the student how to draw on the external angles and measure them and find their sum. Question 2 requires discussion rather than written work. Encourage the student to communicate using the correct mathematical terms, illustrating the discussion with diagrams.

PLENARY ACTIVITY

- **10 facts, 2 minutes; 5 minutes**
  Ask the student to write down 10 facts they can remember about angles in two minutes. Discuss their answers.

HOMEWORK ACTIVITY

- **Time to revise; 45 minutes; page 123**
  Full instructions are given on the activity sheet.

SUPPORT IDEA

- **Using angle properties** Encourage the student to mark every angle they know rather than just trying to find the missing angles. You may wish to support the student by providing 'fact cards' of the angle properties they can use – illustrate these with diagrams.

EXTENSION IDEA

- **Polygons** Extend question 2a) by asking the student to find a formula that connects the number of sides of a polygon to the sum of the internal angles.

PROGRESS AND OBSERVATIONS

STARTER ACTIVITY: COLOUR IT IN

TIMING: 5 MINS

LEARNING OBJECTIVES

- Use conventional terms and notation

EQUIPMENT

- coloured pencils
- protractor

1. On the shape below, colour:

a) side AB red

b) a pair of parallel sides blue

c) a right angle orange

d) angle BCD green

e) a pair of perpendicular sides yellow

f) an acute angle pink

g) an obtuse angle purple.

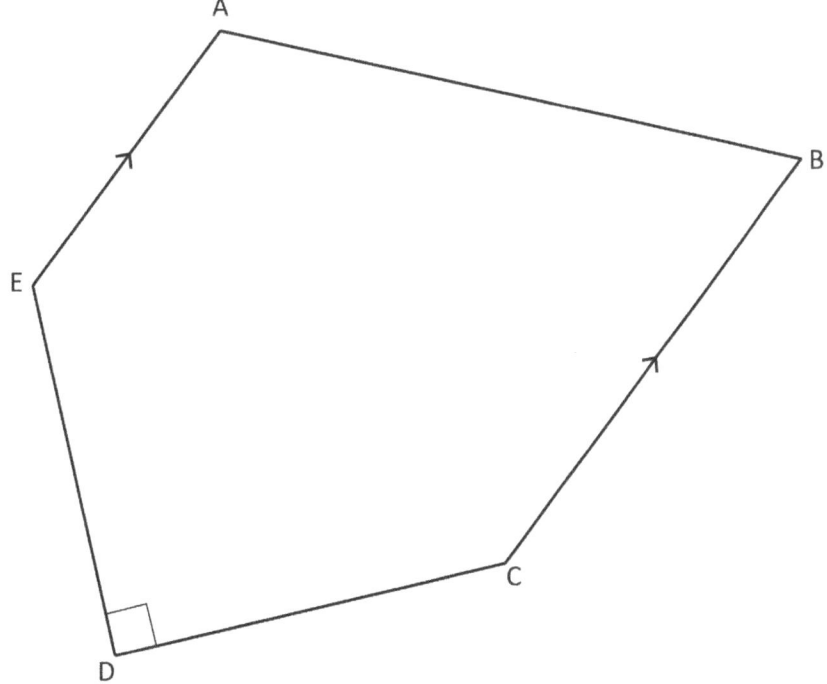

2. What is the name of the shape?

3. Measure the angles. Which is the largest angle?

MAIN ACTIVITY: USING ANGLE PROPERTIES   TIMING: **15** MINS

LEARNING OBJECTIVES   EQUIPMENT

- Apply the properties of angles at a point, on a line, vertically opposite, corresponding and alternate

1. **Calculate the size of each of the labelled angles. For each angle, explain your reasoning. The diagram isn't drawn to scale, so you'll need to use angle rules rather than a protractor!**

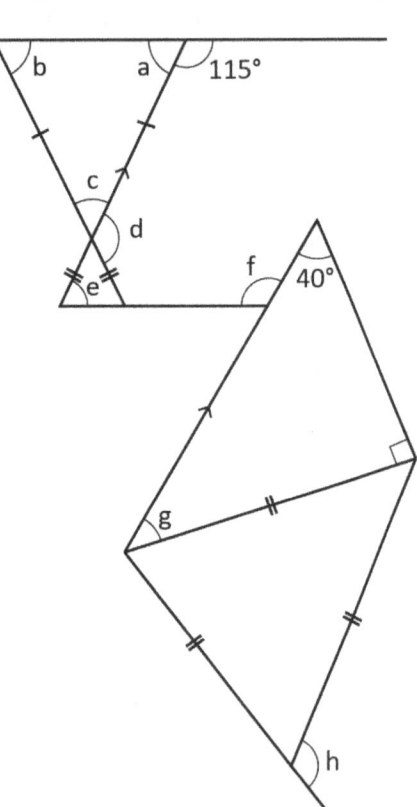

a = ...

b = ...

c = ...

d = ...

e = ...

f = ...

g = ...

h = ...

2. **The triangle opposite has three angles labelled *a*, *b* and *c*.**

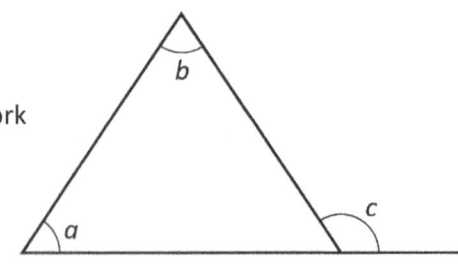

a) Choose two angles smaller than 90° for *a* and *b*, and use them to work out the size of angle *c*.

....

b) Does *a* + *b* = *c*?   ....

c) Try this for other values of *a* and *b*. What do you notice? Can you explain your observation?

....

....

MATHS
— FOUNDATION —

MAIN ACTIVITY: POLYGONS

TIMING: 25 MINS

LEARNING OBJECTIVES	EQUIPMENT
• Calculate and use the sum of angles in polygons	• ruler • protractor

1. Amelia draws a quadrilateral and splits it into two triangles by drawing a line between a pair of opposite vertices.

   Since the angles in a triangle add up to 180°, Amelia works out that the angles in a quadrilateral add up to 360°.

   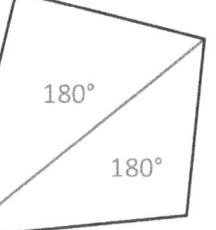

   Use Amelia's method to work out the sum of angles in the polygons in the table below.

   Measure the exterior angles to work out the sum of the exterior angles of a polygon.

Name of shape	Number of sides	Sum of interior angles	Sum of exterior angles
triangle	3	180°	360°
quadrilateral	4	360°	
pentagon	5		
hexagon	6		
heptagon	7		
octagon	8		
nonagon	9		
decagon	10		

2. Discuss the answers to each of these questions with your tutor.

   a) How could you work out the sum of the interior angles of a polygon with *n* sides?

   ---

   b) How could you work out the sum of the exterior angles of a polygon with *n* sides?

   ---

   c) What is the size of one of the external angles of a regular pentagon?

   d) What is the size of one of the internal angles of a regular nonagon?

   e) Can regular octagons and squares tessellate together? Give a reason for your answer.

   ---

   f) Can pentagons and nonagons tessellate together? Give a reason for your answer.

HOMEWORK ACTIVITY: TIME TO REVISE

TIMING: 45 MINS

LEARNING OBJECTIVES

- Apply the properties of angles at a point, on a line, vertically opposite, corresponding and alternate
- Calculate and use the sum of angles in polygons

EQUIPMENT

- index cards/large piece of paper
- coloured pens
- ruler

 1. There are lots of facts you will need to remember about angles for the exam.

Make a set of flash cards or a large poster listing all the facts you have covered in this lesson.

You must include facts about:

- angles on a straight line

- angles round a point

- angles in a triangle

- vertically opposite angles

- corresponding angles

- alternate angles

- co-interior angles

- angles in a quadrilateral

- angles in a polygon

- sum of exterior angles of a polygon

- angles in a regular polygon.

18 ANSWERS

STARTER ACTIVITY: COLOUR IT IN

1. Check student's diagram.   2. pentagon   3. A = 115°, B = 65°, C = 140°, D = 90°, E = 130°; angle C is largest

MAIN ACTIVITY: USING ANGLE PROPERTIES

1. a = 65° (angles on a straight line sum to 180°)

b = 65° (the two angles at the base of the equal sides in an isosceles triangle are equal)

c = 50° (angles in a triangle sum to 180°)

d = 130° (angles on a straight line sum to 180°)

e = 65° (angles on a straight line sum to 180° *or* vertically opposite angles are equal, and the two angles at the base of the equal sides in an isosceles triangle are equal.)

f = 115° (co-interior angles sum to 180 °)

g = 50° (angles in a triangle sum to 180°)

h = 120° (angles in an equilateral triangle are all 60°, and angles on a straight line sum to 180°)

2. a)–c) $a + b = c$ because the sum of the angles in a triangle = 180°, so $c = 180° − (180° − (a + b))$

MAIN ACTIVITY: POLYGONS

1.

Name of shape	Number of sides	Sum of interior angles	Sum of exterior angles
triangle	3	180°	360°
quadrilateral	4	360°	360°
pentagon	5	540°	360°
hexagon	6	720 °	360°
heptagon	7	900 °	360°
octagon	8	1080 °	360°
nonagon	9	1260 °	360°
decagon	10	1440 °	360°

2. a) $(n − 2) × 180°$ or two less than the number of sides multiplied by 180°

b) The external angles of all polygons sum to 360°.

c) 360° ÷ 5 = 72°

d) 1260° ÷ 9 = 140°

e) The internal angle of a regular octagon is 135°. You can put two octagons together with one square at a vertex, since 135 + 135 + 90 = 360° (see diagram).

f) The internal angle of a pentagon is 108°, and the internal angle of a nonagon is 140°. No combination of these will add up to 360°.

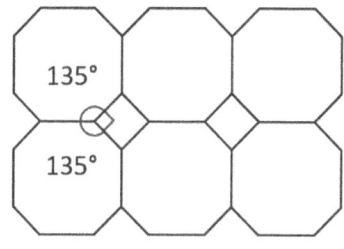

HOMEWORK ACTIVITY: TIME TO REVISE

1. Check student's flashcards or poster.

GLOSSARY

Parallel

Two lines are parallel if they remain the same distance apart. Parallel lines never meet.

Perpendicular

Two lines are perpendicular if the angle between them is 90°.

19 GEOMETRY AND MEASURES: RULER AND COMPASS CONSTRUCTION

LEARNING OBJECTIVES

- Use standard ruler and compass construction
- Construct figures and solve loci problems

SPECIFICATION LINKS

- G1, G2

STARTER ACTIVITY

- **Types of triangle; 5 minutes; page 126**
  Full instructions are given on the activity sheet.

MAIN ACTIVITIES

- **Construction; 25 minutes; page 127**
  For question 1, ensure the student knows how to construct all five constructions. For question 2, explain to the student that in an exam they will usually be given the first construction line.

- **Locus problems; 15 minutes; page 128**
  Explain to the student that in loci diagrams, a solid line means the line is within the area, and a dotted line means the line is not included in the area.

PLENARY ACTIVITY

- **Five top tips; 5 minutes**
  Ask the student to come up with five top tips for construction and loci that they would give to another GCSE maths student.

HOMEWORK ACTIVITY

- **Construction vlogger; 60 minutes; page 129**
  Full instructions are given on the activity sheet. You may wish to ask the student to email the file to you, or they could show you the video on their mobile phone. If the student does not have access to a video recorder (such as a mobile phone or tablet), or if they do not want to record a video, they could write a script for the video instead.

SUPPORT IDEA

- **Construction** Draw the first line of any constructions that the student struggles with to get them started.

EXTENSION IDEA

- **Locus problems** Ask the student to design their own locus problems that will result in particular shapes (e.g. a semicircle, a straight line).

PROGRESS AND OBSERVATIONS

STARTER ACTIVITY: TYPES OF TRIANGLE

TIMING: 5 MINS

LEARNING OBJECTIVES

- Recall and use the properties and definitions of special types of triangle

EQUIPMENT

- ruler

1. On the dotted grid below, all the dots are 1 cm from the dots either side and above or below them.

   Join three dots to make each of these triangles. Which one can't be drawn by joining dots?
   a) an isosceles triangle
   b) a right-angled triangle
   c) an equilateral triangle
   d) a scalene triangle
   e) a triangle that is a combination of more than one of these types of triangle.

MAIN ACTIVITY: CONSTRUCTION

TIMING: 25 MINS

LEARNING OBJECTIVES

- Use standard ruler and compass construction

EQUIPMENT

- pair of compasses
- ruler
- plain paper

1. On a separate piece of paper, draw these constructions.

a) a perpendicular bisector of a line
b) a perpendicular from a point to a line
c) an angle bisector
d) an angle of 90°
e) an angle of 45°

2. Try these exam-style questions.

a) The base of triangle ABC is a line 10 cm long. Draw a line of 10 cm in the middle of a piece of A4 paper and label it AB. Given that <BAC = 45° and <ABC = 90°, use a pair of compasses and a ruler to construct triangle ABC. You must show all your construction lines.

(3 marks)

b) In this garden, a path is laid that bisects angle EDG. Mark the path on the scale diagram. Show all your construction lines.

(2 marks)

Main activity: Locus problems

Timing: 15 mins

1. Points A and B below are two radio transmitters. The two transmitters have a range of up to and including 5 km. The diagram has a scale of 1 cm : 1 km. Mark on the locus of points where signals from both transmitters can be received.

●
A

●
B

2. A dog is tied by a 4 m rope to the corner of a shed. Mark on the locus of points where the dog can roam, using the scale 1 m = 0.5 cm.

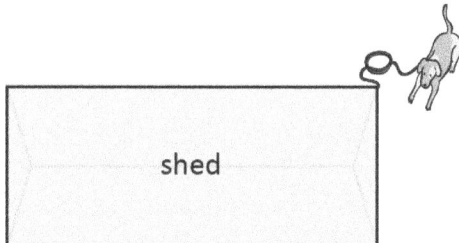
shed

3. A point moves such that it is always more than 2 m from the line AB.
Using a scale of 1 m : 1 cm, mark the locus of the point on the diagram.

●
A

●
B

128

HOMEWORK ACTIVITY: CONSTRUCTION VLOGGER   TIMING: 60 MINS

LEARNING OBJECTIVES
- Use standard ruler and compass construction
- Construct figures and solve loci problems

EQUIPMENT
- technology with filming capacity (phone, tablet, PC, laptop)
- pair of compasses
- ruler

1. **You are going to become a 'construction vlogger'. Make a video or write a script to show other GCSE maths students how to construct the following:**

    a) a perpendicular bisector of a line
    b) an angle bisector
    c) a perpendicular from a point to a line
    d) a right angle
    e) a 45° angle

If you need some help, look at videos that are already available online.

Make sure you bring your video to your next tutorial. You could email it to your tutor or bring it on your phone.

19 ANSWERS

STARTER ACTIVITY: TYPES OF TRIANGLE

1. a)–d) Check student's triangles. It is not possible to make an equilateral triangle by joining dots. There are many possible isosceles, right-angled and scalene triangles.

e) The following 'combination' triangles are possible: scalene and right-angled; isosceles and right-angled.

MAIN ACTIVITY: CONSTRUCTION

1. Check student's constructions.

2. Answers not to scale

a)
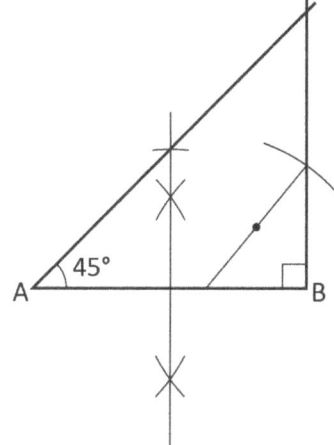

b)
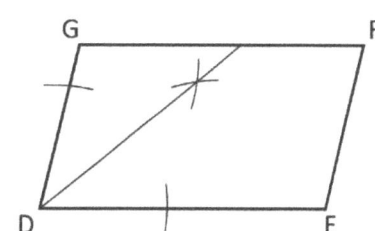

MAIN ACTIVITY: LOCUS PROBLEMS

Answers not to scale

1.
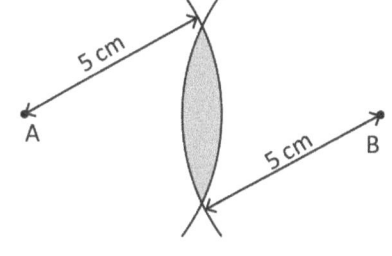

2.
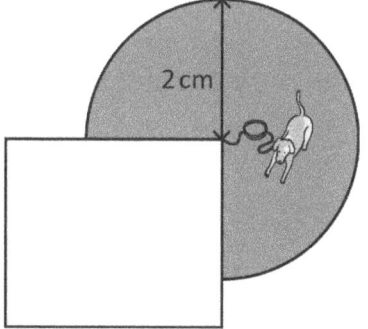

3.
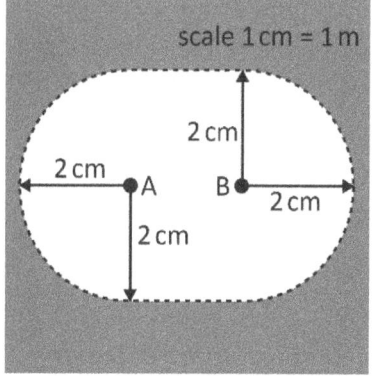

HOMEWORK ACTIVITY: CONSTRUCTION VLOGGER

1. Review student's video and provide feedback as appropriate.

GLOSSARY

Locus (plural loci)

A region or set of points that satisfies particular criteria

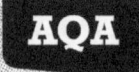  

20 GEOMETRY AND MEASURES: CIRCLES

LEARNING OBJECTIVES

- Identify and apply circle definitions and properties
- Know and use the formulae for area and circumference of a circle
- Calculate arc lengths, angles and areas of sectors of circles

SPECIFICATION LINKS

- G9, G17, G18, N8, A2

STARTER ACTIVITY

- **Parts of a circle; 5 minutes; page 132**
  This activity tests the student's understanding of terminology relating to circles.

MAIN ACTIVITIES

- **Area and circumference; 20 minutes; page 133**
  Ensure that the student is familiar with π and remind them that this is just a number. Remind the student how to give exact solutions, and how to round to a given degree of accuracy. Work through the activity.
- **Parts of circles; 20 minutes; page 134**
  Encourage the student to think logically about what is known and how they can use this to deduce the lengths and areas of arcs and sectors.

PLENARY ACTIVITY

- **Area or circumference?; 5 minutes**
  Give the student different measurements with units (cm/m/mm/inches and cm^2/m^2/mm^2/inches^2) and ask them to identify whether each value is an area or a circumference. Move on to the formulae $2\pi r$ and πr^2. Encourage them to notice that areas all contain the radius squared.

HOMEWORK ACTIVITY

- **Problem solving with circles; 40 minutes; page 135**
  Full instructions are given on the activity sheet.

SUPPORT IDEA

- **Area and circumference** Copy the following table and ask the student to complete it:

Radius (cm)	1	2	3	4	5	6	7	8	9	10
Circumference										
Area										

EXTENSION IDEA

- **Parts of circles** Ask the student to write a formula for finding the area and circumference of semicircles and quarter-circles. Extend this to any size sector.

PROGRESS AND OBSERVATIONS

STARTER ACTIVITY: PARTS OF A CIRCLE

TIMING: 5 MINS

LEARNING OBJECTIVES

- Identify and apply circle definitions and properties

EQUIPMENT

- coloured pens/pencils

1. **On the circle:**

   a) colour the circumference red

   b) draw on a blue radius

   c) draw on an orange diameter

   d) draw on a green chord.

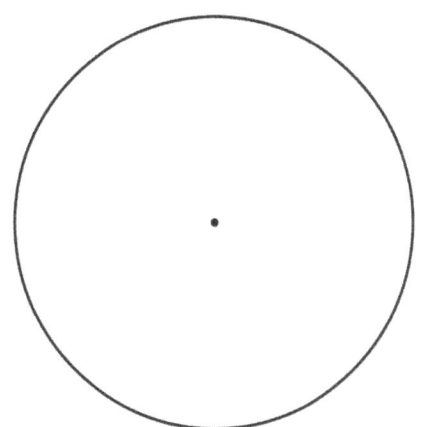

2. **Draw lines to match the words with the correct diagrams.**

   sector   segment   arc   tangent

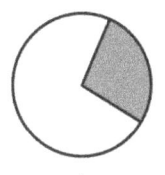   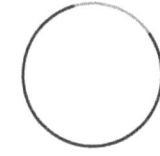 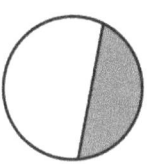 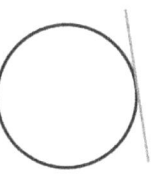

   A   B   C   D

AQA MATHS
— FOUNDATION —

TUTORS' GUILD

MAIN ACTIVITY: AREA AND CIRCUMFERENCE	TIMING: 20 MINS

LEARNING OBJECTIVES

- Know and use the formulae for area and circumference of a circle

EQUIPMENT

- calculator

You need to remember these formulae:

circumference of a circle = $2\pi r$

area of a circle = πr^2

 1. **Work out the circumference and the area of each of these circles. Give your answers to 2 decimal places.**

a)

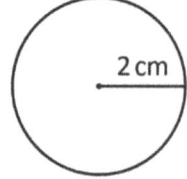

2 cm

circumference = ...

area = ...

b)

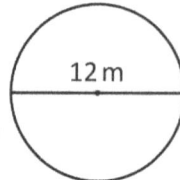

12 m

circumference = ...

area = ...

c) diameter = 15 mm

circumference = ...

area = ...

d) radius = 3.5 cm

circumference = ...

area = ...

 2. **Without using a calculator, write down the exact area and circumference of each of these circles.**

a) diameter = 14 cm ....

b) radius = 0.5 m ....

 3. **A circle has circumference $k\pi$. If the circle has radius 4.5, what is the value of k?**

....

 4. **Work out the radii of these circles to 2 significant figures.**

a) circumference = 100 cm ....

b) area = 25 cm^2 ....

MAIN ACTIVITY: PARTS OF CIRCLES　　　　**TIMING: 20 MINS**

LEARNING OBJECTIVES

- Calculate arc lengths, angles and areas of sectors of circles

EQUIPMENT

1. **How would you work out:**

   a) the area of any semicircle

   ----

   b) the length of an arc on a semicircle

   ----

   c) the perimeter of any semicircle?

   ----

2. **Each of these circles has a diameter of 10 cm. Work out the area of the shaded areas.**
   Give exact answers.

   a)
   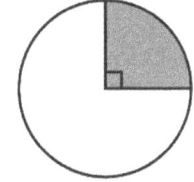

   b)
   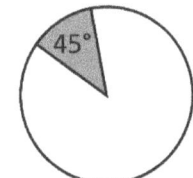

   c)
   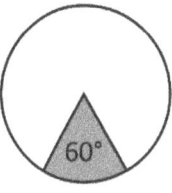

   ....　　........................　　........................

3. **Work out the perimeter of this sector. Give an exact answer.**

   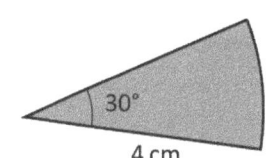

   ....

HOMEWORK ACTIVITY: PROBLEM SOLVING WITH CIRCLES   TIMING: 40 MINS

LEARNING OBJECTIVES

- Identify and apply circle definitions and properties
- Know and use the formulae for area and circumference of a circle
- Calculate arc lengths, angles and areas of sectors of circles

EQUIPMENT

- calculator

1. **A bicycle wheel has a diameter of 65 cm. How many complete revolutions will it make on a 2 km bike ride?**

2. **A stained glass window is made in the following design.**

   a) Giving your answers to 2 decimal places, work out the area of:

       i)   speckled glass   ....

       ii)   grey glass.   ....

   b) Speckled glass costs £15.75 per square metre.
   Grey glass costs £12.00 per square metre.
   What is the cost of glazing the window?

   0.5 m

   0.25 m

3. **A company's logo is in the shape of a square with a semicircle on top.**
   The logo is 15 mm wide.

   a) Work out the total area of the logo. Give your answer to 1 decimal place.

   b) What percentage of the logo is grey? Give your answer to 2 significant figures.

   15 mm

4. **The perimeter of this quarter circle is $m + n\pi$. Work out the values of m and n.**

   8 cm

20 ANSWERS

STARTER ACTIVITY: PARTS OF A CIRCLE

1. Check student's diagram.
2. A sector, B arc, C segment, D tangent

MAIN ACTIVITY: AREA AND CIRCUMFERENCE

1. a) circumference = 12.57 cm   area = 12.57 cm^2 b) circumference = 37.70 m   area = 113.10 m^2
c) circumference = 47.12 mm   area = 176.71 mm^2 d) circumference = 21.99 cm   area = 38.48 cm^2
2. a) area = 49π m^2, circumference = 14π m

b) area = $\dfrac{1}{4}\pi$ cm^2, circumference = π cm

3. $k = 9$
4. a) 16 cm   b) 2.8 cm

MAIN ACTIVITY: PARTS OF CIRCLES

1. a) Find the area of the whole circle and then halve it.
b) Find the circumference of the whole circle and halve it.
c) Find the circumference of the whole circle, halve it, and then add the diameter.

2. a) $\dfrac{25\pi}{4}$ cm   b) $\dfrac{25\pi}{8}$ cm   c) $\dfrac{25\pi}{6}$ cm

3. $\dfrac{2\pi}{3} + 8$ cm

HOMEWORK ACTIVITY: PROBLEM SOLVING WITH CIRCLES

1. 979
2. a) i) 0.79 m^2 ii) 0.98 m^2 b) £24.20
3. a) 313.4 mm^2 b) 28%
4. $m = 16$, $n = 4$

GLOSSARY

Arc
Part of the circumference of a circle

Sector
Part of a circle enclosed by an arc and two radii

Segment
Part of a circle enclosed by a chord and an arc

21 GEOMETRY AND MEASURES: PYTHAGORAS' THEOREM

LEARNING OBJECTIVES

- Construct triangles given three sides
- Know the formula for Pythagoras' theorem and apply it to find lengths in right-angled triangles

SPECIFICATION LINKS

- G20, N3, N6, N7, N8

STARTER ACTIVITY

- **Triangle construction; 5 minutes; page 138**
  Remind the student how to construct a triangle given three sides.

MAIN ACTIVITIES

- **Introducing Pythagoras' theorem; 20 minutes; page 139**
  Model how to find the hypotenuse in a right-angled triangle using Pythagoras' theorem. Ask the student to work through question 1. Then look at the worked example and discuss how to find the length of one of the shorter sides of a triangle. Ask the student to work through question 2. Explain that a number can be left in surd form if you wish to find the exact solution. Show students how to do this for question 1a).
- **Applying Pythagoras' theorem; 20 minutes; page 140**
  Full instructions are given on the activity sheet. To differentiate this activity, you may wish to include extra steps to support the student's reasoning.

PLENARY ACTIVITY

- **Approximate solution; 5 minutes**
  Draw a right-angled triangle with the two shorter sides both labelled 3cm. Invite the student to estimate the length of the longest side. Check by measuring.

HOMEWORK ACTIVITY

- **Investigating Pythagoras; 60 minutes; page 141**
  Full instructions are given on the activity sheet.

SUPPORT IDEA

- **Introducing Pythagoras' theorem** Before starting this activity, draw a 3, 4, 5 right-angled triangle and draw out the square on each side, illustrating the principal before introducing the formula.

EXTENSION IDEA

- **Applying Pythagoras' theorem** Ask the student to explain how to find the length of a line joining any two points on the coordinate grid.

PROGRESS AND OBSERVATIONS

STARTER ACTIVITY: TRIANGLE CONSTRUCTION

TIMING: 5 MINS

LEARNING OBJECTIVES

- Construct triangles given three sides

EQUIPMENT

- pair of compasses
- ruler
- protractor

1. Construct the following triangles in the space below.

   a) AB = 5 cm, BC = 3 cm, AC = 4 cm

   b) AB = 5 cm, BC = 13 cm, AC = 12 cm

   c) AB = 6 cm, BC = 8 cm, AC = 10 cm

2. What do you notice about all these triangles?

MAIN ACTIVITY: INTRODUCING PYTHAGORAS' THEOREM   TIMING: 20 MINS

LEARNING OBJECTIVES

- Know the formula for Pythagoras' theorem and apply it to find lengths in right-angled triangles

EQUIPMENT

- calculator

Pythagoras' theorem states that in a right-angled triangle, $a^2 + b^2 = c^2$ where c is the hypotenuse.

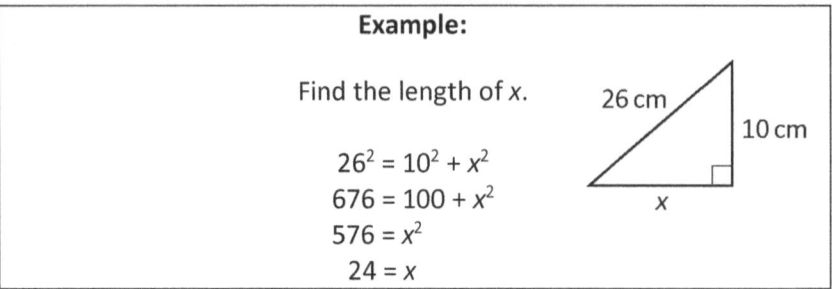

Example:

Find the length of x.

$26^2 = 10^2 + x^2$
$676 = 100 + x^2$
$576 = x^2$
$24 = x$

26 cm   10 cm x

1. **Work out the length of the hypotenuse for each of these triangles.**
   Give your answers to 2 significant figures.

a)

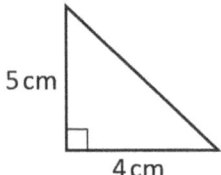

5 cm
4 cm

....

b)

0.9 m
12 cm

....

c)

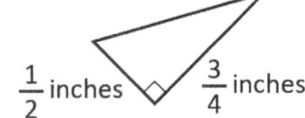

$\frac{1}{2}$ inches $\frac{3}{4}$ inches

....

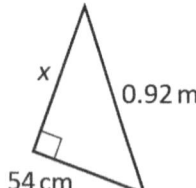

2. **Calculate the length of the side labelled x for each of these triangles.**
   Give your answers to 2 significant figures.

a)

5 cm
x
21 mm

....

b)

x 0.92 m
54 cm

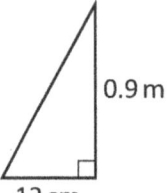

....

MATHS
— FOUNDATION —

AQA

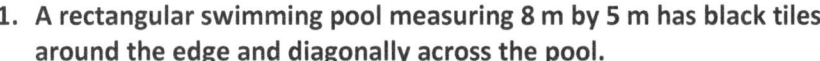

| MAIN ACTIVITY: APPLYING PYTHAGORAS' THEOREM | TIMING: **20** MINS |

LEARNING OBJECTIVES

- Know the formula for Pythagoras' theorem and apply it to find lengths in right-angled triangles

EQUIPMENT

- calculator

1. **A rectangular swimming pool measuring 8 m by 5 m has black tiles around the edge and diagonally across the pool.**

   The tiles can be bought in strips 1 m long and cost £12.99 per metre. Work out the cost of the decorative tiling for the pool.
   Show all your workings.

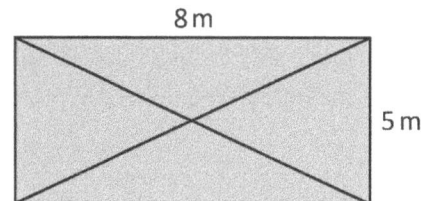

2. **The points A (3, 6) and B (−1, −3) are plotted on the coordinate grid and joined as shown.**
   Work out the exact length of the line AB.

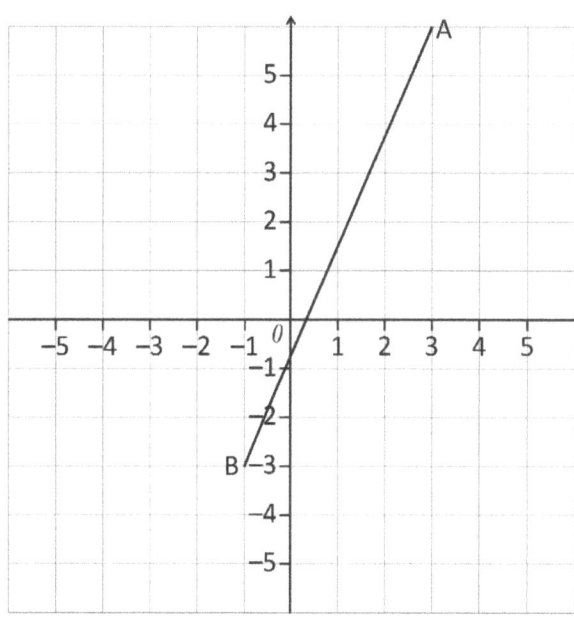

3. **The dimensions of three triangles are given below.**
   Which are right-angled? Give a reason for your answer.

   a)   AB = 7 cm, BC = 25 cm, AC = 24 cm

   b)   AB = 0.3 m, BC = 0.2 m, AC = 0.4 m

   c)   AB = 17 cm, BC = 15 cm, AC = 8 cm

MATHS
— FOUNDATION —

HOMEWORK ACTIVITY: INVESTIGATING PYTHAGORAS   **TIMING: 60 MINS**

LEARNING OBJECTIVES
- Know the formula for Pythagoras' theorem and apply it to find lengths in right-angled triangles

EQUIPMENT
- calculator
- dice

1. Look at this triangle.

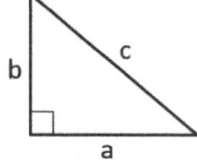

   a) Roll a dice twice to give the side lengths a and b for this triangle. Now work out the length of c.

   b) Can you find values of a and b which give a whole number value for c? Is there more than one pair of values that give a whole number for c?

   c) Now roll the dice twice to give the values of a and c. Use the larger value for c. If the values are the same, roll again. Work out the length of b.

   d) Can you find values of a and c which give a whole number value for b?

2. A Pythagorean triplet is a set of three whole numbers which are the lengths of three sides of a right-angled triangle.

   a) Decide if the following sets of three numbers are Pythagorean triplets:

      i)   5, 12, 13   --

      ii) 10, 24, 26   --

      iii) 15, 36, 39   --

   b) Use any patterns you notice to find two more Pythagorean triplets.

21 ANSWERS

STARTER ACTIVITY: TRIANGLE CONSTRUCTION

1. Check student's triangles.

2. All the triangles are right-angled triangles.

MAIN ACTIVITY: INTRODUCING PYTHAGORAS' THEOREM

1. a) 6.4 cm   b) 91 cm or 9.1 m   c) 0.90 inches

2. a) 45 mm or 4.5 cm   b) 74 cm or 0.74 m

MAIN ACTIVITY: APPLYING PYTHAGORAS' THEOREM

1. a) £584.55

2. $\sqrt{97}$

3. a) and c) because the sum of the squares of the shorter sides is equal to the square of the longer side.

HOMEWORK ACTIVITY: INVESTIGATING PYTHAGORAS

1. a)–c) Check student's answers using the table of possible values of c:

		a or b					
		1	2	3	4	5	6
a or b	1	1.4					
	2	2.2	2.8				
	3	3.2	3.6	4.2			
	4	4.1	4.5	5	5.7		
	5	5.1	5.4	5.8	6.4	7.1	
	6	6.1	6.3	6.7	7.2	7.8	8.5

d) There is only one whole-number combination: when a and b are 3 and 4, c is 5.

2. a) i), ii) and iii) are all Pythagorean triplets.

b) Any two triplets of the form 5a, 12a, 13a.

GLOSSARY

Pythagorean triplet

Three positive integers a, b and c which satisfy $a^2 + b^2 = c^2$

22 GEOMETRY AND MEASURES: USING TRIGONOMETRY

LEARNING OBJECTIVES

- Rearrange a simple equation to make a given variable the subject
- Know and use the formulae for the trigonometric ratios and apply them to find angles and lengths in right-angled triangles in 2D figures

SPECIFICATION LINKS

- G1, G20

STARTER ACTIVITY

- **Rearranging an equation; 5 minutes; page 144**
  This activity is a precursor to rearranging the trigonometric ratios. Encourage the student to think about the triangle that relates the three values in the equation, if necessary writing the equation in a formula triangle with 12 at the top.

MAIN ACTIVITIES

- **Finding missing lengths; 20 minutes; page 145**
  Work through the worked example with the student. You may wish to invent a mnemonic to use to help the student remember SOHCAHTOA (e.g. Studying On Holiday Can Always Have Two Obvious Advantages). Before the student tackles question 3, introduce the terms 'angle of elevation' and 'angle of depression'.
- **Finding missing angles; 20 minutes; page 146**
  Work through the example with the student. Emphasise that the system is very similar to finding the lengths of sides. Make sure they know how to use their calculator to find $\sin^{-1}$, $\cos^{-1}$ and $\tan^{-1}$.

PLENARY ACTIVITY

- **Flow chart; 5 minutes**
  Design flow charts to show how to calculate a) a missing side and b) a missing angle in a right-angled triangle.

HOMEWORK ACTIVITY

- **Solving problems; 60 minutes; page 147**
  Full instructions are given on the activity sheet.

SUPPORT IDEA

- **Finding missing lengths** If the student can identify which trigonometric ratio to use, but struggles to rearrange it, it might help to display SOHCAHTOA in three formula triangles like this:

$$\frac{O}{S \quad H} \qquad \frac{A}{C \quad H} \qquad \frac{O}{T \quad A}$$

They can then cover the side they wish to calculate in the relevant triangle and see what calculation they need to do.

EXTENSION IDEA

- **Finding missing lengths** Sketch a cuboid and invite students to find the lengths of sides/diagonals, given appropriate lengths and angles. To extend further, challenge them to find the angle between a side and a diagonal.

PROGRESS AND OBSERVATIONS

STARTER ACTIVITY: REARRANGING AN EQUATION

TIMING: 5 MINS

LEARNING OBJECTIVES

- Rearrange a simple equation to make a given variable the subject

EQUIPMENT

We know that $4 = \dfrac{12}{3}$.

This equation can be rearranged to give: $4 \times 3 = 12$ or $3 = \dfrac{12}{4}$.

1. Use this method to make x the subject of each of these equations.

a) $a = \dfrac{x}{b}$

b) $a = \dfrac{b}{x}$

AQA MATHS
— FOUNDATION —

MAIN ACTIVITY: FINDING MISSING LENGTHS   **TIMING: 20 MINS**

LEARNING OBJECTIVES

- Know and use the formulae for the trigonometric ratios and apply them to find lengths in right-angled triangles in 2D figures

EQUIPMENT

- calculator

Example:

Find the length of side x in this triangle.
Give your answer to 2 decimal places.

Label the sides: adjacent, opposite and hypotenuse.
Use SOHCAHTOA to decide whether to use sin, cos or tan.
Substitute the values you know into the relevant formula and solve:

$$\cos 40 = \frac{\text{adjacent}}{12}$$

$$= 9.19 \text{ cm}$$

1. Find the length of y in the triangle above. Give your answer to 2 decimal places.

....

2. Calculate the length of the side AB in each of the triangles below. Give your answers to 2 decimal places.

a)

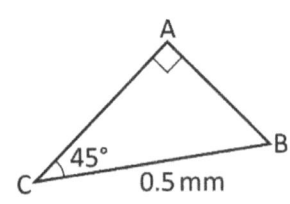

AB = ...

b)

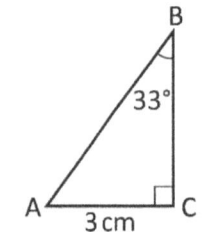

AB = ...

c)

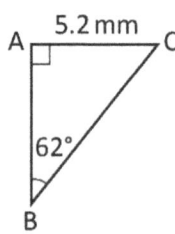

AB = ...

3. A man is watching a bird in a tree. The angle of elevation from the horizontal is 35°.
He is watching the bird from the ground 12 m horizontally from the bottom of the tree.

How far up the tree is the bird?
Give your answer to the nearest 10 cm.

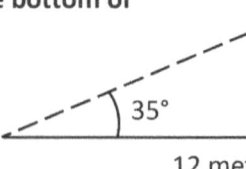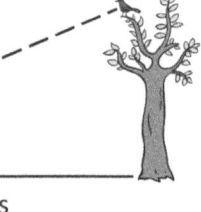

12 metres

Remember, angle of elevation means looking up, and angle of depression means looking down.

....

....

MAIN ACTIVITY: FINDING MISSING ANGLES

TIMING: 20 MINS

LEARNING OBJECTIVES

- Know and use the formulae for the trigonometric ratios and apply them to find lengths in right angled triangles in 2D figures

EQUIPMENT

- calculator

Example:

Find angle x in this triangle.
Give your answer to 2 decimal places.

Label the sides: adjacent, opposite and hypotenuse.
Use SOHCAHTOA to decide whether to use sin, cos or tan.
Substitute the values you know into the relevant formula and solve:

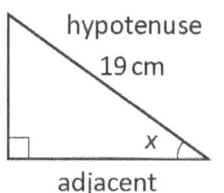

$$\sin x = \frac{10}{19}$$

$$x = \sin^{-1}\left(\frac{10}{19}\right)$$

$$x = 31.76°$$

---

1. **Calculate the size of angle ABC in each of the triangles below.**

a)

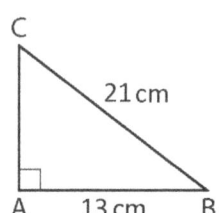

angle ABC = ...

b)

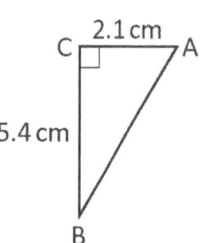

angle ABC = ...

c)

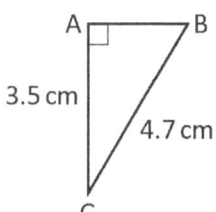

angle ABC = ...

d)

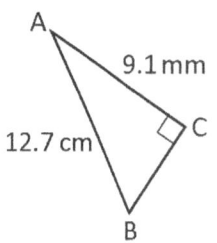

angle ABC = ...

2. **A helicopter is hovering, waiting to land. It is 20 feet above ground level and 100 feet horizontally from the landing point.**

   Calculate the angle of depression from the helicopter to the landing point.

   ....

HOMEWORK ACTIVITY: SOLVING PROBLEMS

TIMING: 60 MINS

LEARNING OBJECTIVES
- Know and use the formulae for the trigonometric ratios and apply them to find angles and lengths in right-angled triangles in 2D figures

EQUIPMENT
- calculator
- ruler
- protractor

1. Sophie is using the trigonometric ratios to find the size of angle *x*. She has written:

   tan *x* = 12 × 18
   x = 89.7°

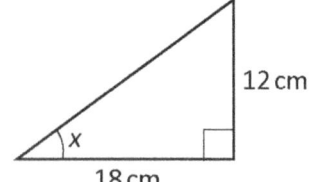

   a) Explain what Sophie has done wrong.

   ....

   ....

   b) Work out the correct value of *x*.   ....

2. A wheelchair access ramp is 2 m long. The step is 50 cm high. Work out the angle the ramp is to the horizontal.

   ....

   ....

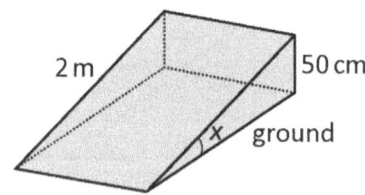

3. Use trigonometry to calculate the lengths of sides *a* and *b* in this triangle.

   ....

   ....

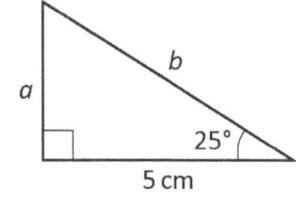

4. A roof is designed so that the angle of the slope with the horizontal is 45°. Work out the lengths of the sloping sides.

   ....

   ....

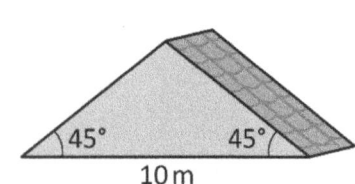

5. Draw three right-angled triangles below. Label one side and one angle (you can either measure or make up these values). Use trigonometry to work out the lengths of the other two sides in each triangle.

22 Answers

Starter activity: Rearranging an equation

1. a) $x = a \times b$ b) $x = \dfrac{b}{a}$

Main activity: Finding missing lengths

1. 7.71 cm
2. a) 0.35 mm   b) 5.51 cm   c) 2.76 mm
3. 8.4 m

Main activity: Finding missing angles

1. a) 51.8°   b) 21.3°   c) 48.1°   d) 4.1°
2. 11.3°

Homework activity: Solving problems

1. a) She has multiplied instead of dividing.   b) 33.7°
2. 14.5°
3. a = 2.3 cm, b = 5.5 cm
4. both sides = 7.1 m
5. Student's own answers

Glossary

Angle of elevation
The angle formed between the horizontal and a straight line which makes an angle above the horizontal

Angle of depression
The angle formed between the horizontal and a straight line which makes an angle below the horizontal

23 GEOMETRY AND MEASURES: TRIGONOMETRIC VALUES

LEARNING OBJECTIVES	SPECIFICATION LINKS

LEARNING OBJECTIVES

- Know the exact values of sin θ and cos θ for θ = 0°, 30°, 45°, 60° and 90°
- Know the exact values of tan θ for θ = 0°, 30°, 45° and 60°
- Order numbers including fractions, decimals and surds

SPECIFICATION LINKS

- G20, G21, N1

STARTER ACTIVITY

- **Ordering numbers; 5 minutes; page 150**
  Remind the student that an exact value can be given by leaving the number in surd form. Encourage the student to estimate the value of each surd (knowing that $\sqrt{1} = 1$ and $\sqrt{4} = 2$ tells you that $\sqrt{2}$ and $\sqrt{3}$ must be between 1 and 2).

MAIN ACTIVITIES

- **Special triangles; 25 minutes; page 151**
  For question 3, you may wish to explain how values of sin 0, cos 0, sin 90 and cos 90 occur by sketching graphs of the functions.
- **Using the trigonometric values; 15 minutes; page 152**
  The student may find it helpful to refer to the table from question 3 of the previous activity.

PLENARY ACTIVITY

- **What's my angle?; 5 minutes**
  Sketch two right-angled triangles, one with sides of length 1 cm, 1 cm and $\sqrt{2}$ cm and one with sides of length 2 cm, 1 cm and $\sqrt{5}$ cm. Ask the student to label the angles of the triangles with their sizes.

HOMEWORK ACTIVITY

- **Trigonometry revision; 60 minutes; page 153**
  It is important for the student to learn these trigonometric ratios, so the homework activities are aimed at creating revision resources. If the student already has their own method for revising, you may wish to change these activities to suit them.

SUPPORT IDEAS

- **Special triangles** Before starting this activity, revisit SOHCAHTOA. Remind the student how to find missing sides and angles.
- **Using the trigonometric values** Before starting this activity, explain to the student that if they are asked to work out a missing angle or side in a right-angled triangle *without* a calculator, they can assume it will be one of these special triangles.

EXTENSION IDEA

- **Using the trigonometric values** Ask the student to design a problem of their own for you to answer.

PROGRESS AND OBSERVATIONS

STARTER ACTIVITY: ORDERING NUMBERS   TIMING: 5 MINS

LEARNING OBJECTIVES	EQUIPMENT
• Order numbers including fractions, decimals and surds	none

1. Write these numbers in ascending order:

$\frac{1}{2}$ $\sqrt{2}$ 1 5^2

....

2. Write these numbers in descending order:

$\frac{3}{4}$ $\sqrt{3}$ -1.7 4.9

....

MATHS
— FOUNDATION —

MAIN ACTIVITY: SPECIAL TRIANGLES

TIMING: 25 MINS

LEARNING OBJECTIVES

- Know the exact values of sin θ and cos θ for θ = 0°, 30°, 45°, 60° and 90°
- Know the exact values of tan θ for θ = 0°, 30°, 45° and 60°

EQUIPMENT

- calculator

1. **Triangle ABC is an isosceles right-angled triangle with two sides of length 1.**

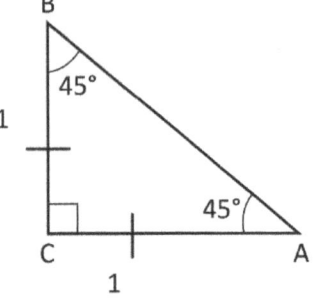

   a) Use Pythagoras' theorem to work out the length of AB. Leave your answer in surd form. Label the diagram with your answer.

   b) Use SOHCAHTOA to work out the values of:

      i) sin 45° ....

      ii) cos 45° ....

      iii) tan 45° ....

2. **Triangle DEF is an equilateral triangle with sides of length 2 cm. The point G is the midpoint of DE.**

   a) What are the sizes of angles D, E and F?

      D = ...   E = ...   F = ...

   b) Draw a vertical line joining F with point G. What is the length of line DG?

      ....

   c) Use Pythagoras' theorem to work out the length of FG.   ....

   d) Write the size of the angles in triangle DGF.   ....

   e) Use triangle DGF to write down the exact values of:

      i) sin 30° ....   ii) sin 60° ....

      ii) cos 30° ....   iv) cos 60° ....

      v) tan 30° ....   vi) tan 60° ....

3. **Use your answers to questions 1 and 2 and your calculator to complete this table.**

θ	0°	30°	45°	60°	90°
sin θ					
cos θ					
tan θ					

MAIN ACTIVITY: USING THE TRIGONOMETRIC VALUES   TIMING: 15 MINS

LEARNING OBJECTIVES	EQUIPMENT
• Know the exact values of sin θ and cos θ for θ = 0°, 30°, 45°, 60° and 90°	none
• Know the exact values of tan θ for θ = 0°, 30°, 45° and 60°	

1. Write these values in order, starting with the smallest: sin 30°, tan 45°, cos 90°

....

2. Work out the size of angle *x* in this triangle.

....

3. ABCD is a square. The diagonal BD is $\sqrt{2}$ cm long.
   Work out each of the following. Give reasons for your answers.

   a)   the length of side DC

   ....

   b)   angle BDC

   ....

   c)   angle DBC

   ....

4. PQR is a right-angled triangle.

   a)   Write down the sizes of angles PQR and PRQ.
        Give reasons for your answers.

   PQR = ...   PRQ = ...

   ....

   b)   Work out the length of side PQ.

   ....

HOMEWORK ACTIVITY: TRIGONOMETRY REVISION   TIMING: 60 MINS

LEARNING OBJECTIVES

- Know the exact values of sin θ and cos θ for θ = 0°, 30°, 45°, 60° and 90°
- Know the exact values of tan θ for θ = 0°, 30°, 45° and 60°

EQUIPMENT

- large paper
- scissors

 1. Draw a poster or flow chart showing how to use SOHCAHTOA to find unknown angles and unknown lengths.

 2. Cut out these cards and use them to learn the trigonometric values. You could play snap or pairs, display them on the wall, or ask someone to test you on them.

sin 45°	$\frac{1}{\sqrt{2}}$	cos 60°	$\frac{1}{2}$
cos 45°	$\frac{1}{\sqrt{2}}$	tan 60°	$\sqrt{3}$
tan 45°	1	sin 0°	0
sin 30°	$\frac{1}{2}$	cos 0°	1
cos 30°	$\frac{\sqrt{3}}{2}$	tan 0°	0
tan 30°	$\frac{1}{\sqrt{3}}$	sin 90°	1
sin 60°	$\frac{\sqrt{3}}{2}$	cos 90°	0

23 ANSWERS

STARTER ACTIVITY: ORDERING NUMBERS

1. $\frac{1}{2}$ 1 $\sqrt{2}$ 5^2

2. 4.9 $\sqrt{3}$ $\frac{3}{4}$ -1.7

MAIN ACTIVITY: SPECIAL TRIANGLES

1. a) $\sqrt{2}$ b) i) $\sin 45 = \frac{1}{\sqrt{2}}$ ii) $\cos 45 = \frac{1}{\sqrt{2}}$ iii) $\tan 45 = 1$

2. a) D, E and F = 60°   b) Check FG has been drawn correctly. Length of line DG = 1 cm.

c) $\sqrt{3}$ d) 60° 30° and 90°

e) i) $\sin 30 = \frac{1}{2}$ ii) $\sin 60 = \frac{\sqrt{3}}{2}$ iii) $\cos 30 = \frac{\sqrt{3}}{2}$

iv) $\cos 60 = \frac{1}{2}$ v) $\tan 30 = \frac{1}{\sqrt{3}}$ vi) $\tan 60 = \sqrt{3}$

3.

θ	0°	30°	45°	60°	90°
$\sin \theta$	0	$\frac{1}{2}$	$\frac{1}{\sqrt{2}}$	$\frac{\sqrt{3}}{2}$	1
$\cos \theta$	1	$\frac{\sqrt{3}}{2}$	$\frac{1}{\sqrt{2}}$	$\frac{1}{2}$	0
$\tan \theta$	0	$\frac{1}{\sqrt{3}}$	1	$\sqrt{3}$	

MAIN ACTIVITY: USING THE TRIGONOMETRIC VALUES

1. cos 90, sin 30, tan 45
2. $x = 30°$
3. Check student's reasoning.
a) side DC = 1 cm   b) angle BDC = 45°   c) angle DBC = 45°
4. Check student's reasoning.
a) PQR = 60°   PRQ = 30°   b) side PQ = 1 cm

HOMEWORK ACTIVITY: TRIGONOMETRY REVISION

1–2. Check student's work. You could test the student on the trigonometric values.

24 GEOMETRY AND MEASURES: PROPERTIES OF SHAPES

LEARNING OBJECTIVES

- Derive and apply the properties of special types of quadrilaterals and triangle and other plane figures
- Know the properties of types of triangle
- Recall and use the properties and definitions of special types of quadrilateral

SPECIFICATION LINKS

- G1, G4, G6, G20, A8, A9, A17, A21

STARTER ACTIVITY

- **Properties of a triangle; 5 minutes; page 156**
  You may wish to encourage the student to sketch the triangles, marking on any equal sides or angles.

MAIN ACTIVITIES

- **Quadrilaterals; 25 minutes; page 157**
  Ask the student to complete the table of information about quadrilaterals. They should fill in the columns as follows:
  Sketch it! – the student should sketch the shape using a ruler and mark on any equal sides, parallel sides or right angles using appropriate notation.
  Sides – the student should indicate any special properties the quadrilateral has (e.g. all sides are equal).
  Angles – the student should indicate any special angle properties particular to that quadrilateral (e.g. pairs of equal angles).
  Diagonals – the student should indicate any special properties of the diagonals (e.g. bisect/intersect at right angles).
  Lines of symmetry – the student should indicate how many lines of symmetry the shape has.
  Sum of internal angles – the student should recognise that this is always 360°.
- **Using the properties of quadrilaterals; 15 minutes; page 158**
  Full instructions are given on the activity sheet.

PLENARY ACTIVITY

- **How many?; 5 minutes**
  Ask the student to decide how many quadrilaterals there are that:
  a) contain a right angle   b) have at least one pair of parallel sides   c) contain an obtuse angle.

HOMEWORK ACTIVITY

- **Quadrilaterals on coordinate grids; 30 minutes; page 159**
  Full instructions are given on the activity sheet.

SUPPORT IDEA

- **Quadrilaterals** Describe a quadrilateral to the student and challenge them to sketch it. Then swap roles.

EXTENSION IDEA

- **Quadrilaterals** Draw up a similar table and test the student on non-quadrilateral 2D shapes.

PROGRESS AND OBSERVATIONS

STARTER ACTIVITY: PROPERTIES OF A TRIANGLE   TIMING: 5 MINS

LEARNING OBJECTIVES

* Know the properties of types of triangle

EQUIPMENT

1. **Draw lines to match each statement to at least one of the types of triangle below. Some statements will apply to more than one type of triangle.**

A   all angles are equal

B   two angles are equal

C   no angles are equal

D   may contain a right angle

E   all sides have equal length

F   two sides have equal length

G   no sides have equal length

H   has no lines of symmetry

I   has one line of symmetry

J   has three lines of symmetry

K   the sum of the internal angles is 180°

| scalene triangle |

| isosceles triangle |

| equilateral triangle |

MAIN ACTIVITY: QUADRILATERALS

TIMING: 25 MINS

LEARNING OBJECTIVES
- Recall and use the properties and definitions of special types of quadrilateral

EQUIPMENT
- ruler

 1. Complete this table to show information about some different types of quadrilateral.

Name of shape	Sketch it!	Sides	Angles	Diagonals	Lines of symmetry	Sum of internal angles
square						
rectangle						
rhombus						
parallelogram						
trapezium						
kite						

MAIN ACTIVITY: USING THE PROPERTIES OF QUADRILATERALS TIMING: 15 MINS

LEARNING OBJECTIVES	EQUIPMENT
• Recall and use the properties and definitions of special types of quadrilateral	none

1. Decide if these statements are true or false. Explain each of your answers to your tutor.

   a) All quadrilaterals contain at least one right angle.   ....

   b) A square is also a rhombus.   ....

   c) At least two of the angles in a rhombus must be obtuse.   ....

   d) A parallelogram must contain two obtuse angles.   ....

2. List all the quadrilaterals whose diagonals are perpendicular.

   ....

   ....

3. Quadrilateral ABCD has four sides of equal length. Angle ABC is twice the size of angle BCD.

   a) What is the name of the quadrilateral?   ....

   b) Work out the size of angle ABC.   ....

4. Look at this drawing.

   a) Form and solve an equation to calculate the value of x.

   ....

   ....

   b) Use your answer to part a) to work out the size of angle ABE.

   ....

   ....

The drawing shows a quadrilateral with vertices labelled. Angle at E is $2x°$, angle at D is $x + 20°$, angle at B is $x + 30°$, and angle at C is $2x + 10°$. Points A, B, C lie on a line at the bottom.

HOMEWORK ACTIVITY: QUADRILATERALS ON COORDINATE GRIDS TIMING: **30** MINS

LEARNING OBJECTIVES

- Derive and apply the properties of special types of quadrilateral

EQUIPMENT

- ruler
- index cards

1. **Here is a blank coordinate grid.**

   a) On the coordinate grid, plot and label these points:
   A (2, 6), B (−4, 3) and C (−5, −1).

   b) ABCD is a parallelogram. Mark on the point D and write down the coordinates of D.

   ....

   c) Work out the length of line AB.
   Give your answer to 1 decimal place.

   ....

   d) Work out the length of line BC.
   Give your answer to 1 decimal place.

   ....

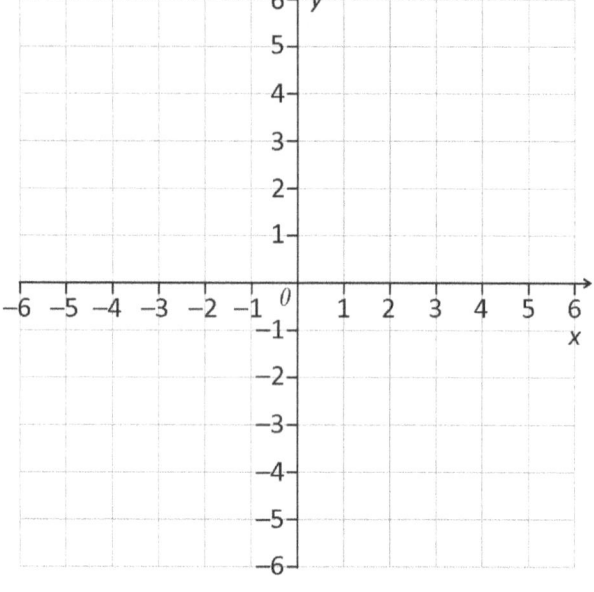

2. **The points A, B and C are three of the vertices of a kite.**

   a) Write down four possible pairs of coordinates for the vertex D.

   ....

   b) Draw a straight line on the grid indicating the possible positions of D.

   c) Work out the equation of the line going through these points.

   ....

3. **Make a set of revision cards showing six different quadrilaterals and listing all their properties.**

24 ANSWERS

STARTER ACTIVITY: PROPERTIES OF A TRIANGLE

1. scalene triangle: C, D, G, H, K   isosceles triangle: B, D, F, I, K   equilateral triangle: A, E, J, K

MAIN ACTIVITY: QUADRILATERALS

1. Check student's sketches.

Name of shape	Sides	Angles	Diagonals	Lines of symmetry	Sum of internal angles
square	All sides are equal and opposite sides are parallel.	All angles are 90°.	Diagonals bisect at right angle and are equal in length.	4	360°
rectangle	Two pairs of sides are equal and opposite sides are parallel.	All angles are 90°.	Diagonals bisect and are equal in length.	2	360°
rhombus	All sides are equal and opposite sides are parallel.	Diagonally opposite angles are equal.	Diagonals bisect at right angles.	2	360°
parallelogram	Two pairs of sides are equal and opposite sides are parallel.	Diagonally opposite angles are equal.	Diagonals bisect.	0	360°
trapezium	One pair of opposite sides is parallel.			0 or 1 if isosceles trapezium	360°
kite	Two pairs of sides are equal.	One pair of diagonally opposite angles is equal.	Diagonals are perpendicular; one diagonal is bisected.	1	360°

MAIN ACTIVITY: USING THE PROPERTIES OF QUADRILATERALS

1. a) False – any sketch of a quadrilateral without a right angle will justify this.
b) True – a square has all the properties of a rhombus.
c) False – a square is a type of rhombus and has four right angles.
d) False – a rectangle is a parallelogram and has four right angles.
2. square, rhombus, kite
3. a) rhombus   b) 120°
4. a) $6x + 60 = 360$, $x = 50°$ b) 100°

HOMEWORK ACTIVITY: QUADRILATERALS ON COORDINATE GRIDS

1. a) Check student's graph.   b) (1, 2)   c) 6.7   d) 4.1
2. a) Student's own answers   b) Check line shows $y = x + 2$.   c) $y = x + 2$
3. Student's own work

GLOSSARY

Diagonal of a quadrilateral
The straight line joining opposite vertices

25 GEOMETRY AND MEASURES: CONGRUENT AND SIMILAR SHAPES

LEARNING OBJECTIVES	SPECIFICATION LINKS

LEARNING OBJECTIVES

- Identify right-angled triangles using Pythagoras' theorem
- Use the basic congruence criteria for triangles (SSS, SAS, ASA, RHS)
- Apply angle facts, triangle congruence, similarity and properties of quadrilaterals to conjecture and derive results about angles and sides, and obtain simple proofs

SPECIFICATION LINKS

- G1, G2, G3, G5, G6, G19

STARTER ACTIVITY

- **Am I right-angled?; 5 minutes; page 162**
  If necessary, recap Pythagoras' theorem before starting the activity. Full instructions for the activity are given on the sheet.

MAIN ACTIVITIES

- **Congruence criteria; 25 minutes; page 163**
  This activity is designed to familiarise the student with the information needed to check whether triangles meet congruence criteria (SSS, SAS, ASA, RHS). If necessary, remind the student how to construct triangles.
- **Using congruence; 15 minutes; page 164**
  Remind the student of the mathematical meaning of the terms 'congruent' and 'similar'. Ask them to spot some congruent and similar shapes around the room. Establish how to check for similarity in shapes (by checking the ratio between equivalent sides is constant).

PLENARY ACTIVITY

- **Congruent and similar; 5 minutes**
  Ask the student to explain the meaning of the terms 'congruent' and 'similar' and explain how you can identify pairs of congruent/similar shapes.

HOMEWORK ACTIVITY

- **Similarity and congruence; 30 minutes; page 165**
  Before setting the homework, ensure the student understands the definitions of congruence and similarity.

SUPPORT IDEA

- **Using congruence** Draw a set of triangles and give the student some tracing paper. Invite the student to identify congruent pairs of shapes using the tracing paper.

EXTENSION IDEA

- **Congruence criteria** Ask the student *'what information do you need in order to be able to construct a unique triangle?'*

PROGRESS AND OBSERVATIONS

STARTER ACTIVITY: AM I RIGHT-ANGLED?   TIMING: 5 MINS

LEARNING OBJECTIVES

- Identify right-angled triangles using Pythagoras' theorem

EQUIPMENT

- calculator

1. Explain to your tutor how you could work out whether these triangles contain a right angle. They are not drawn to scale.

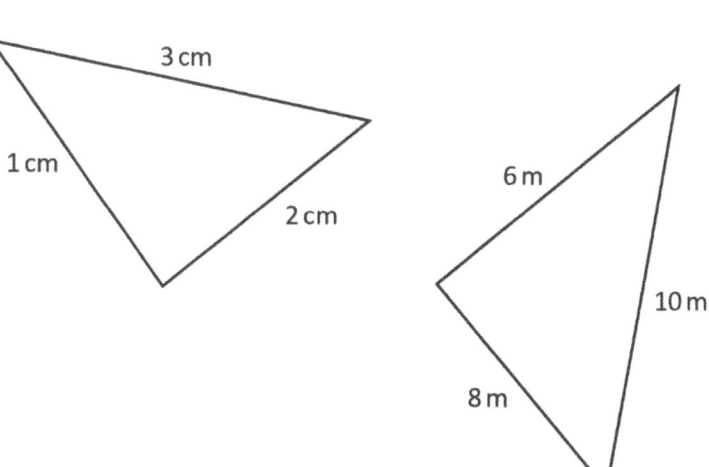

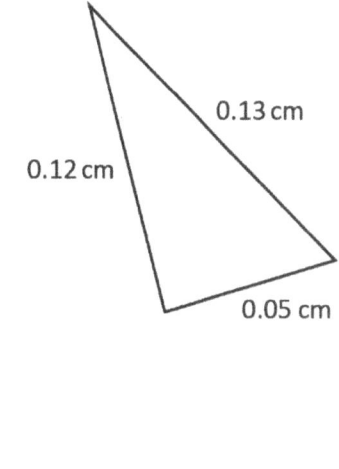

2. If there are any right angles, mark them onto the triangles.

MAIN ACTIVITY: CONGRUENCE CRITERIA   **TIMING: 25 MINS**

LEARNING OBJECTIVES

- Apply angle facts, triangle congruence, similarity and properties of quadrilaterals to conjecture and derive results about angles and sides, and obtain simple proofs

EQUIPMENT

- plain paper
- ruler
- protractor

 1. **On a separate piece of paper, construct the following triangles from these different combinations of information.**

a) three sides (SSS): AB = 10 cm, BC = 5 cm, AC = 7 cm

b) two sides and an included angle (SAS): AB = 6.5 cm, BC = 3.5 cm, <ABC = 75°

c) two angles and an included side (ASA): <ABC = 47°, <ACB = 25°, BC = 8 cm

d) one right angle, the hypotenuse and one other side (RHS): AB = 9 cm, <ABC = 90°, AC = 5 cm

 2. **A question in an exam asks students to construct a triangle with AB = 8 cm, AC = 6.5 cm and <ABC = 50°. Phyllis and Abdul both attempt to construct this triangle.**

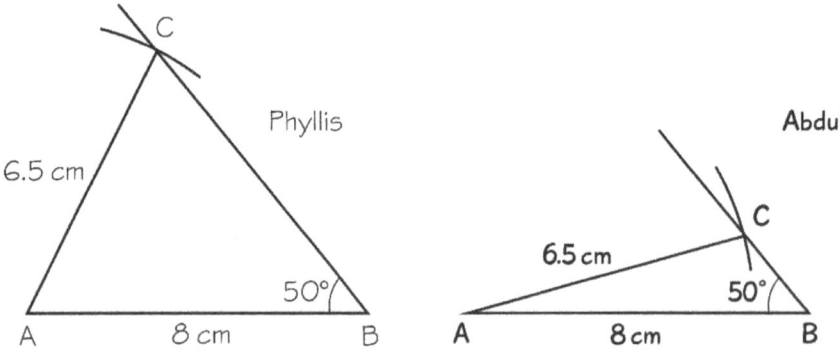

a) Measure the triangles and check their constructions.

b) Are the triangles congruent? Explain your answer.

----

c) Are the triangles similar? Explain your answer.

----

d) Construct two different triangles, both with AB = 9 cm, BC = 7 cm and <BAC = 45°.

MATHS
— FOUNDATION —

MAIN ACTIVITY: USING CONGRUENCE	**TIMING: 15 MINS**

LEARNING OBJECTIVES	EQUIPMENT
• Use the basic congruence criteria for triangles (SSS, SAS, ASA, RHS) • Apply angle facts, triangle congruence, similarity and properties of quadrilaterals to conjecture and derive results about angles and sides, and obtain simple proofs	none

Two shapes are **congruent** if they are identical. All corresponding angles are equal and all corresponding sides are equal.

Two shape are **similar** if all corresponding angles are equal and corresponding sides are in the same ratio. One is an enlargement of the other.

1. **The area of triangle ABC is 32.5 cm². Triangle PQR is congruent to triangle ABC. What is the area of triangle PQR?**

---

2. **ABCD is a rectangle. E is the midpoint of CD. Decide whether these statements are true or false. Explain your answers to your tutor.**

    a) Triangle ABE is isosceles.

    ---

    b) Triangle BCE is congruent to triangle ADE.

    ---

    c) The area of triangle BCE is twice the area of triangle ABE.

    ---

3. **Triangles DEF and XYZ are similar.**

    a) What is the size of angle DFE?

    ---

    b) Work out the length of side XY.

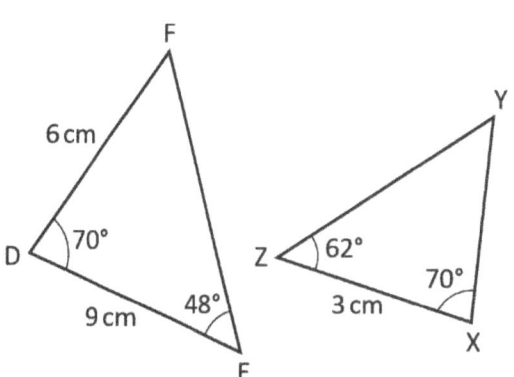

HOMEWORK ACTIVITY: SIMILARITY AND CONGRUENCE   **TIMING: 30 MINS**

LEARNING OBJECTIVES

- Use the basic congruence criteria for triangles
  (SSS, SAS, ASA, RHS)
- Apply angle facts, triangle congruence, similarity and properties
  of quadrilaterals to conjecture and derive results about angles
  and sides, and obtain simple proofs

EQUIPMENT

 1. **From the triangles below, choose a pair that matches each description and explain your answers.**

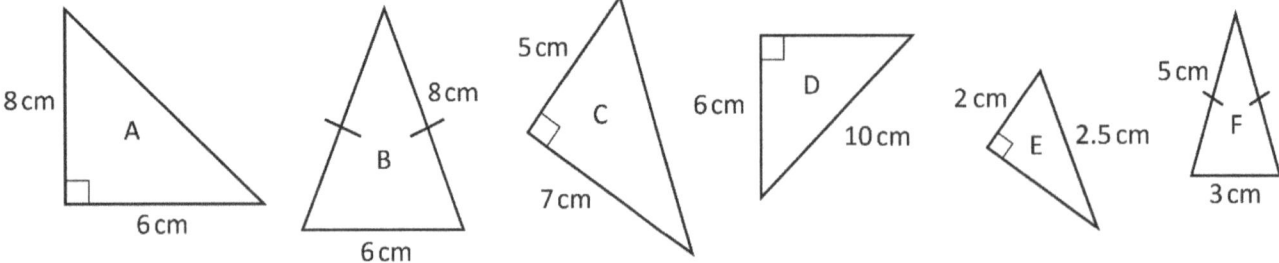

a) congruent

....

b) similar

....

 2. **In the diagram, AC = BC and CD = CE.**

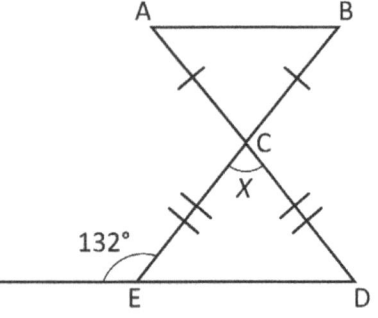

a) Work out the size of angle *x*. Show your working out.

....

b) Explain why triangle ABC is similar to triangle DEC.

....

....

 3. **By drawing one straight line, how many ways can you cut:**

a) a rectangle into two congruent shapes   ....

b) an isosceles triangle into two congruent shapes   ....

c) an equilateral triangle into two congruent shapes   ....

d) a regular octagon into two congruent shapes   ....

e) a regular pentagon into two congruent shapes?   ....

25 ANSWERS

STARTER ACTIVITY: AM I RIGHT-ANGLED?

1. A triangle is right-angled if the sum of the squares of the two shorter sides is equal to the square of the hypotenuse. The first triangle is not right-angled. The second and third are; the right angle is opposite the longest side (hypotenuse).

MAIN ACTIVITY: CONGRUENCE CRITERIA

1. a)–d) Check student's constructions.
2. a) Both triangles are correctly constructed.
b) No – for triangles to be congruent all the sides and all the angles must be equal.
c) No – for triangles to be similar one must be an enlargement of the other, therefore all the angles must be equal and all the sides must be in the same ratio to the equivalent side on the other triangle.
d) Check student's constructions.

MAIN ACTIVITY: USING CONGRUENCE

1. $32.5cm^2$
2. a) True, AE = BE.
b) True, BC = AD, DE = EC and BE = AE.
c) False, area of ABE = $0.5 \times$ AB $\times$ AD, area of BCE = $0.5 \times (0.5 \times$ AB $\times$ AD).
3. a) 62°   b) 4.5 cm

HOMEWORK ACTIVITY: SIMILARITY AND CONGRUENCE

1. a) A and D are congruent since all sides are equal length and all angles are equal.
b) A and E (or D and E) are similar since all angles are equal and corresponding sides are in the same ratio.
2. a) <CED = 180° − 132° = 48°
$x = 180 − 2 \times 48 = 84°$
b) Since AC = BC and CD = CE, AC : DC and BC : EC. ECD = BCA (vertically opposite angles are equal).
So triangles ABC and DEC are similar because two pairs of corresponding sides are in the same ratio and the included angle is equal.
3. a) 4   b) 1   c) 3   d) 8   e) 5

GLOSSARY

Congruent
Congruent shapes are exactly the same shape and size (all corresponding angles are equal and all corresponding sides are equal).

Similar
Similar shapes are the same shape but one is an enlargement of the other (all corresponding angles are equal and corresponding sides are in the same ratio).

26 GEOMETRY AND MEASURES: TRANSFORMATIONS

LEARNING OBJECTIVES

- Apply the properties of special types of quadrilateral
- Identify, describe and construct congruent and similar shapes, including on coordinate axes, by considering rotation, reflection, translation and enlargement (including fractional scale factors)

SPECIFICATION LINKS

- G4, G7

STARTER ACTIVITY

- **Where are the vertices?; 5 minutes; page 168**
  Full instructions are given on the activity sheet.

MAIN ACTIVITIES

- **Reflect, rotate, translate; 20 minutes; page 169**
  Remind the student how to carry out translations (given a vector), rotations and reflections. Model an example on a coordinate grid, stressing the language used and information given. Ask the student to work through question 1. Discuss with the student whether the transformed shapes are congruent or not.
- **Enlargements; 20 minutes; page 170**
  Show the student how to enlarge a shape given the scale factor of enlargement and a centre of enlargement. Encourage them to recognise that the enlarged shapes are similar to the original shape.

PLENARY ACTIVITY

- **What information?; 5 minutes**
  Ask the student to list the information required to carry out:
  a) an enlargement   b) a translation   c) a rotation   d) a reflection.

HOMEWORK ACTIVITY

- **Describing transformations; 40 minutes; page 171**
  Before setting this homework, show the student how to find the centre of enlargement and scale factor of enlargement.

SUPPORT IDEA

- **Enlargements** Ask the student to find the ratio between corresponding sides of the shapes they have drawn and emphasise that this ratio is always the scale factor of enlargement. Discuss how you might calculate what length the sides should be when you are enlarging by different scale factors. For question 2, ask the student to just enlarge the shape by the scale factors given, ignoring the centres of enlargement.

EXTENSION IDEA

- **Reflect, rotate, translate** Invite the student to suggest how you could transform other pairs of triangles onto each other using two transformation (e.g. A to E). Encourage the student to notice there is usually more than one way.

PROGRESS AND OBSERVATIONS

STARTER ACTIVITY: WHERE ARE THE VERTICES?   TIMING: 5 MINS

LEARNING OBJECTIVES

- Apply the properties of special types of quadrilateral

EQUIPMENT

- ruler

1. **Two vertices of a quadrilateral are marked on the coordinate axis below.**

   Write down two possible pairs of coordinates for the other vertices if the shape is:

   a) a square   ....

   b) a parallelogram   ....

   c) a kite   ....

   d) a rhombus.   ....

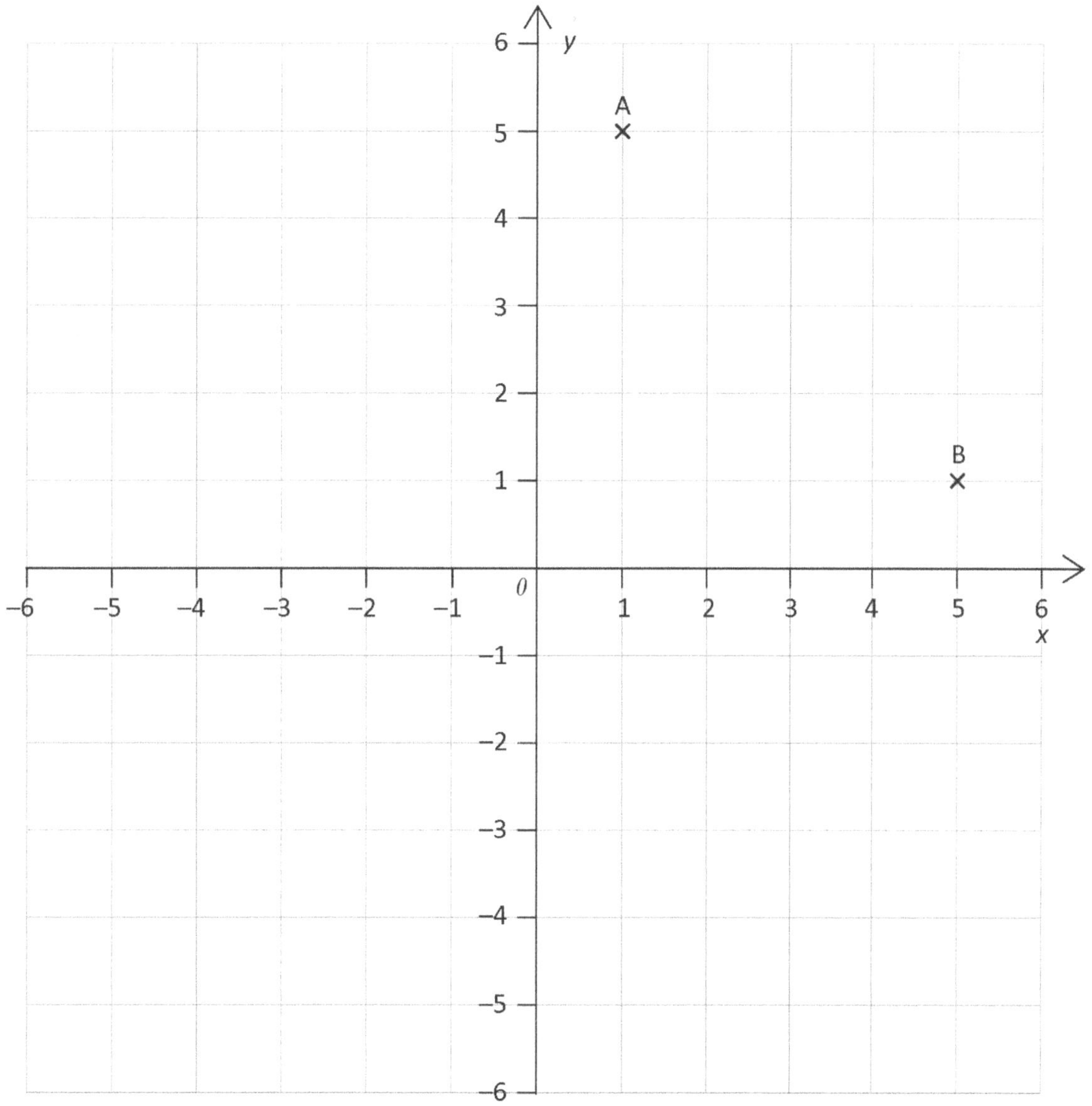

MAIN ACTIVITY: REFLECT, ROTATE, TRANSLATE   **TIMING: 20 MINS**

LEARNING OBJECTIVES

- Identify, describe and construct congruent and similar shapes, including on coordinate axes, by considering rotation, reflection and translation

EQUIPMENT

- ruler
- tracing paper
- sharp pencil

1. **Look at shape A on this coordinate grid.**

   a) Translate shape A through $\begin{pmatrix} -6 \\ 2 \end{pmatrix}$. Label the image B.

   b) Rotate shape B through 90° about the point (−1, 2). Label the image C.

   c) On the axis draw the line $y = -x$. Then reflect triangle C in the line $y = -x$. Label the image D.

   d) Rotate D about the point (−3, −1) through 180°. Label the image E.

   e) Translate shape E through $\begin{pmatrix} 4 \\ -4 \end{pmatrix}$. Label the image F.

   f) Reflect F in the *y*-axis and label the image G.

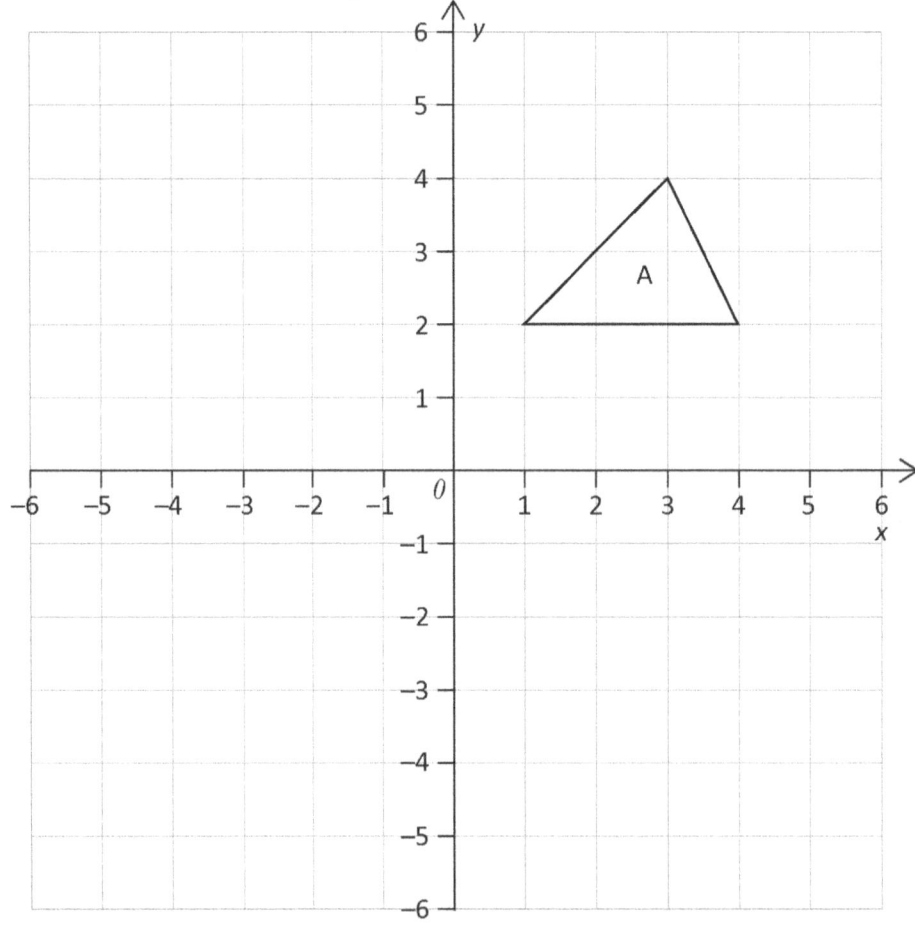

2. **Describe a single transformation that maps:**

   a) G onto B   ....

   b) F onto D   ....

   c) A onto C.   ....

MAIN ACTIVITY: ENLARGEMENTS

TIMING: 20 MINS

LEARNING OBJECTIVES

- Identify, describe and construct congruent and similar shapes, including on coordinate axes, by considering enlargement (including fractional scale factors)

EQUIPMENT

- ruler
- sharp pencil
- squared paper

1. **Draw the shapes described below to scale on squared paper.**

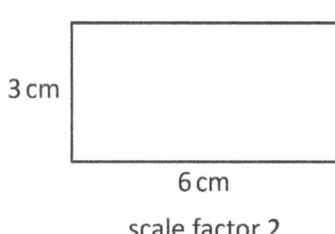

3 cm

6 cm

scale factor 2

4 cm

6 cm

scale factor 3

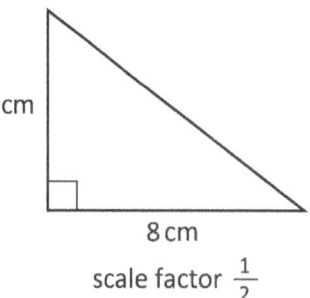

6 cm

8 cm

scale factor $\frac{1}{2}$

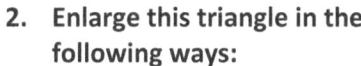

a) Enlarge each shape by the scale factor indicated.
b) Are the pairs of shapes congruent, similar, or neither? Explain your answers to your tutor.

2. **Enlarge this triangle in the following ways:**

a) by scale factor 3
with centre of enlargement A

b) by scale factor 2
with centre of enlargement B

c) by scale factor $\frac{1}{2}$
with centre of enlargement C.

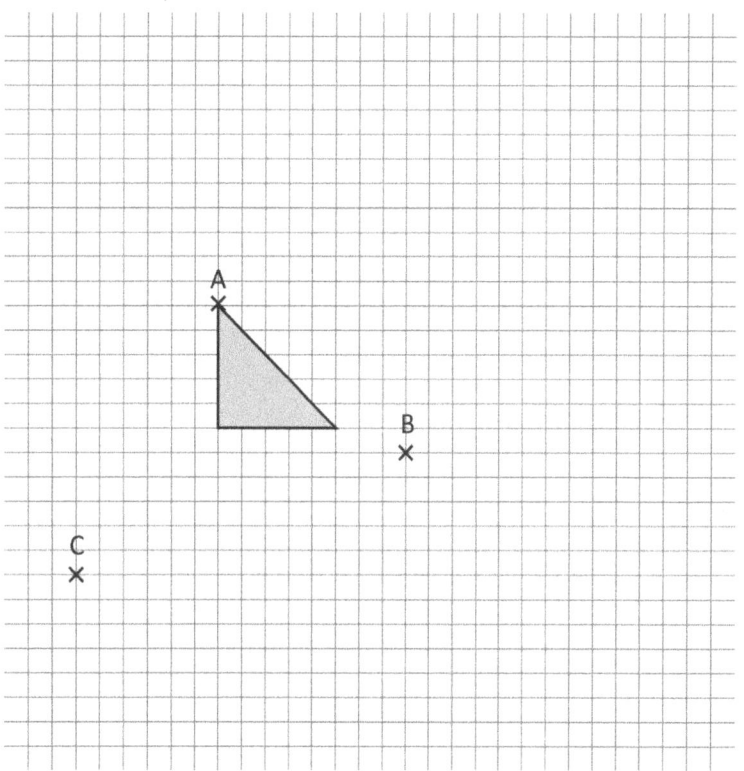

3. **How could you check the size of your enlargements in question 2?**

....

....

MATHS
— FOUNDATION —

HOMEWORK ACTIVITY: DESCRIBING TRANSFORMATIONS   **TIMING: 40 MINS**

LEARNING OBJECTIVES
- Identify, describe and construct congruent and similar shapes, including on coordinate axes, by considering rotation, reflection, translation and enlargement (including fractional scale factors)

EQUIPMENT
- ruler
- sharp pencil
- tracing paper

1. **Describe fully the transformation which maps shape X onto each of these shapes.**

   a) A

      ....

   b) B

      ....

   c) C

      ....

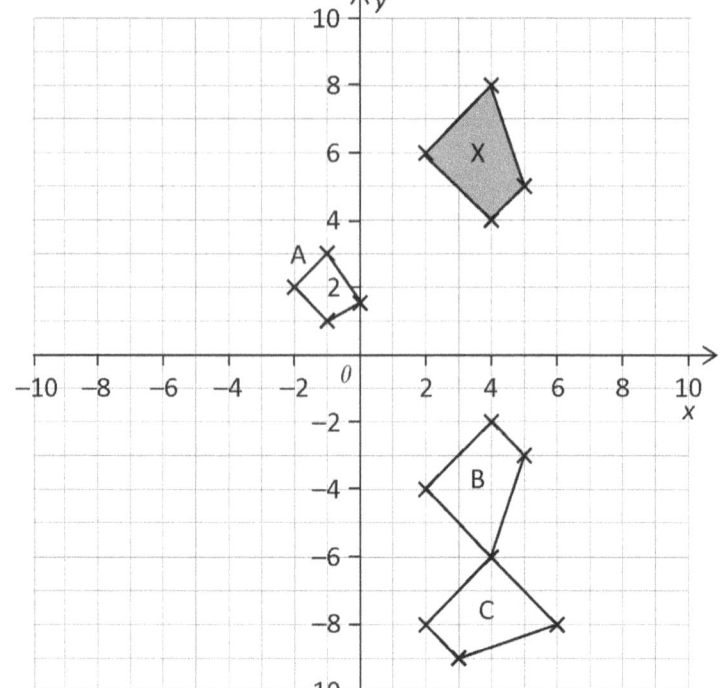

2. **Complete the statements below using the words congruent or similar.**

   a) A shape and its image after a reflection are   ....

   b) A shape and its image after a rotation are   ....

   c) A shape and its image after an enlargement are   ....

   d) A shape and its image after a translation are   ....

3. **Decide if these statements are always true (AT), sometimes true (ST) or never true (NT).**

   a) If a shape is enlarged, its perimeter is twice as long.   ....

   b) If a shape is enlarged, the angles in the image are equal to the angles in the original shape.   ....

   c) If a shape is enlarged, the image is smaller.   ....

   d) If a shape is enlarged, the image is larger.   ....

26 Answers

STARTER ACTIVITY: WHERE ARE THE VERTICES?

1. Check student's answers – there are a number of possible correct answers for each question.

MAIN ACTIVITY: REFLECT, ROTATE, TRANSLATE

1.

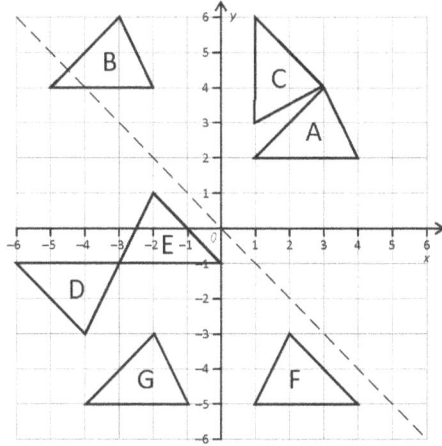

2. a) translation $\begin{pmatrix} -1 \\ 9 \end{pmatrix}$ b) rotation of 180° about (–1, –3)

c) rotation of 90° clockwise (or 270° anticlockwise) about (3, 4)

MAIN ACTIVITY: ENLARGEMENTS

1. a) Check student's answers.
b) similar
2.

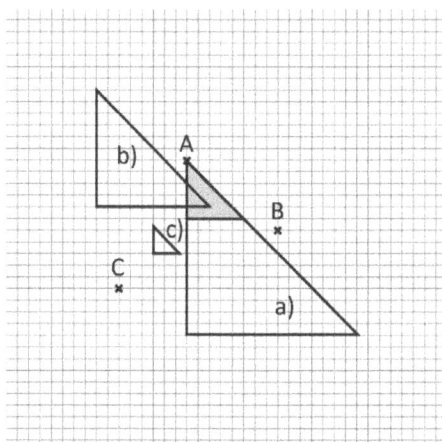

3. You could check them by checking the lengths of the sides have been multiplied by the scale factor of enlargement.

HOMEWORK ACTIVITY: DESCRIBING TRANSFORMATIONS

1. a) enlargement scale factor = $\frac{1}{2}$, centre = (–6, –2)

b) reflection in $y = 1$
c) rotation through 90° clockwise about (–3, –1)
2. a) congruent   b) congruent   c) similar   d) congruent
3. a) ST   b) AT   c) ST   d) ST

27 GEOMETRY AND MEASURES: BEARINGS AND SCALE

LEARNING OBJECTIVES

- Measure line segments and angles in geometric figures
- Interpret maps and scale drawings
- Use bearings

SPECIFICATION LINKS

- G1, G15, R2

STARTER ACTIVITY

- **Estimating and measuring; 5 minutes; page 174**
  Full instructions are given on the activity sheet.

MAIN ACTIVITIES

- **Bearings and scale; 25 minutes; page 175**
  Model how to measure a bearing, and remind the student that they always have three digits. Ask the student to measure the bearings of points A to E from the central point marked with an X. Discuss scale and ask the student to calculate the distance between X and the points A to E in kilometres. Ask the student to mark on points F and G, giving them bearings and distances (in km) from any point on the diagram.

- **Map work; 15 minutes; page 176**
  You may wish to show the student a map of the local area and establish what scale means, or you could discuss how scales work on maps. Work through question 1, establishing when to multiply and when to divide. Agree that distances must be given in sensible units, for example, a distance of 100 000 cm would always be converted into 1 km.

PLENARY ACTIVITY

- **Points on a compass; 5 minutes**
  Draw the four cardinal points on a compass and ask the student to give you the bearing angle between N and E/S/W. Extend to NE, SE, NW, SW.

HOMEWORK ACTIVITY

- **Bingo bearings; 30 minutes; page 177**
  Ask the student to complete a row, a column or the whole board dependent on ability.

SUPPORT IDEA

- **Map work** Look at the scale used on a map of the local area and work out the distance between various different pairs of points. Then invent some imaginary places for the student to mark on the map given the real distance from a fixed point.

EXTENSION IDEA

- **Bearings and scale** Ask the student to find the bearing from other points, e.g. from A to B. Challenge the student to predict the bearing before measuring it.

PROGRESS AND OBSERVATIONS

STARTER ACTIVITY: ESTIMATING AND MEASURING

TIMING: 5 MINS

LEARNING OBJECTIVES

- Measure line segments and angles in geometric figures

EQUIPMENT

- protractor
- ruler

1. Estimate the size of angles *a, b, c, d*. Measure them with a protractor to check your estimates.

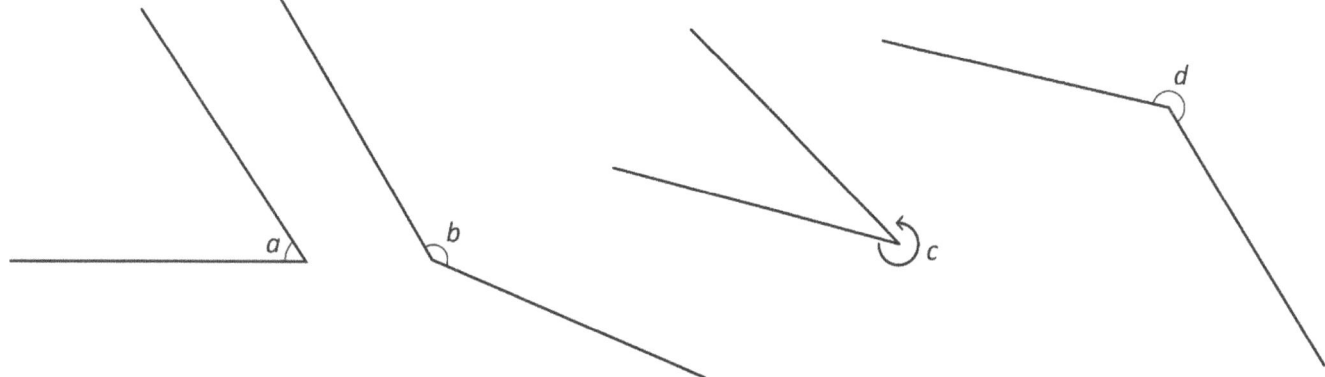

angle *a*	estimate:	measured: ....
angle *b*	estimate:	measured: ....
angle *c*	estimate:	measured: ....
angle *d*	estimate:	measured: ....

2. In the figure ABCD, measure the length of side AB and angle *x*.

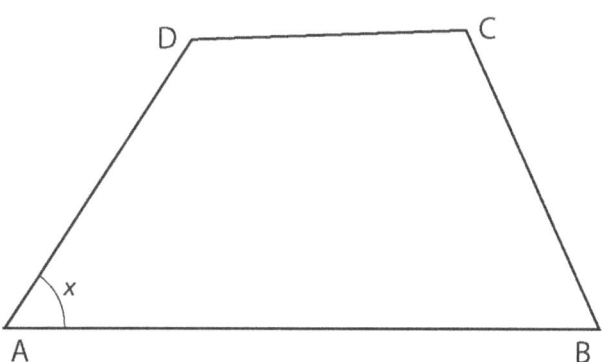

side AB = ...

angle *x* = ...

MAIN ACTIVITY: BEARINGS AND SCALE

TIMING: 25 MINS

LEARNING OBJECTIVES

- Interpret maps and scale drawings
- Use bearings

EQUIPMENT

- protractor
- ruler

1. **Work out the bearings and distances your tutor asks for.**

×
E

1:100 000

×
A

×

×
D

×
B

×
C

MATHS
— FOUNDATION —

AQA

MAIN ACTIVITY: MAP WORK

TIMING: 15 MINS

LEARNING OBJECTIVES

- Interpret maps and scale drawings
- Use bearings

EQUIPMENT

- ruler
- protractor
- pair of compasses

---

1. For maps, a scale is used so that the distance between two points can be calculated. Some maps use the scale 1 : 12 500. This means that 1 cm on the map represents 12 500 cm in real life.

   a) What is 12 500 cm in metres?

   ....

   b) What would 20 cm on the map represent in real life?

   ....

   c) What would 3 km in real life be represented by on the map?

   ....

   d) How do you know whether to multiply or divide when using scales?

   ....

2. Each of the following are scales on a map. Work out what 1 cm on the map represents in real life.

   a) 1 : 1 000 000

   ....

   b) 1 : 10 000

   ....

   c) 1 : 50 000

   ....

   d) 1 : 1000

   ....

3. A ship is on a bearing of 065° from a lighthouse.

   It is 4 km from the lighthouse.

   a) Mark the position of the ship on the map.

   b) A helicopter is flying over the sea on a bearing of 280° from the ship and is 3 km from the lighthouse. Mark the position of the helicopter.

   c) How far, in kilometres, is the helicopter from the ship?

   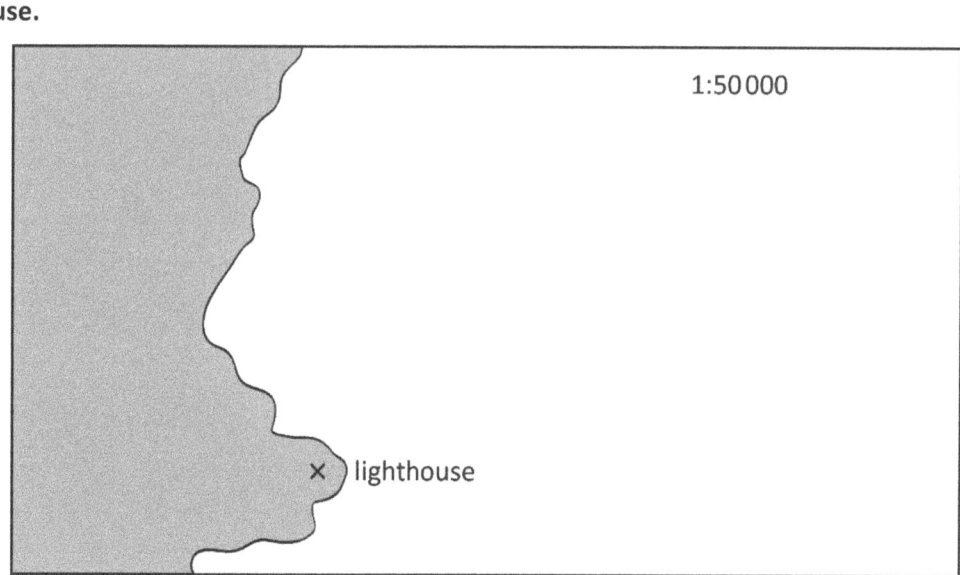

   ....

HOMEWORK ACTIVITY: BINGO BEARINGS   TIMING: 30 MINS

LEARNING OBJECTIVES

- Measure line segments and angles in geometric figures
- Interpret maps and scale drawings
- Use bearings

EQUIPMENT

- ruler
- protractor
- plain paper

1. **Complete this bingo board of questions. Write your answers on a separate piece of paper.**

Estimate the size of < PQR in this triangle. Q P R	Measure the sides of rectangle ABCD and work out the perimeter and area. A B D C	Work out the bearing of B from A. × A × B
Mark on a point C so that <ABC = 135° B A	The scale on a map is 1 : 250 000. What does 1 cm on the map represent?	The bearing of A from B is 120°. What is the bearing of B from A?
The distance between Petersfield and Liphook is 16km. On a map with scale 1 : 200 000 what is the distance between the two towns?	C is on a bearing of 125° from point D. Mark on a possible position for point C. × D	C is on a bearing of 120° from B and 215° from A. Mark the point C on the diagram. × A × B

27 ANSWERS

STARTER ACTIVITY: ESTIMATING AND MEASURING

1. $a = 55°$, $b = 143°$, $c = 330°$, $d = 225°$. Accept answers with a range of ±5°.
2. AB = 7.8 cm, angle $x = 57°$

MAIN ACTIVITY: BEARINGS AND SCALE

1. X to A = 050°, 8.8 km	X to B = 140°, 4.8 km	X to C = 195°, 7.7 km
X to D = 245°, 5.7 km	X to E = 355°, 10.7 km	

MAIN ACTIVITY: MAP WORK

1. a) 125 m   b) 2.5 km   c) 24 cm   d) Student's own answer
2. a) 10 km   b) 100 m   c) 0.5 km (or 500 m)   d) 10 m
3. a) and b)

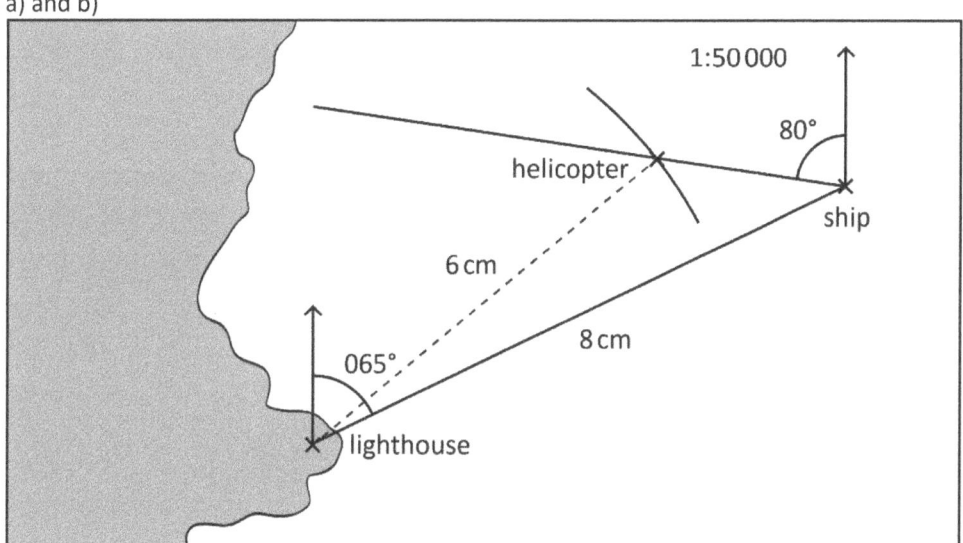

c) 1.35 km

HOMEWORK ACTIVITY: BINGO BEARINGS

1.

80° (± 5°)	perimeter = 11.4 cm area = 7.8 cm²	210°
Check student's angle.	2.5 km	300°
8 cm	Check student's bearing.	Check student's diagram.

GLOSSARY

Bearing

The angle from one point to another, measured clockwise from north. Bearings always have three figures.

28 GEOMETRY AND MEASURES: 3D SHAPES

<table>
<tr><td>

LEARNING OBJECTIVES

- Classify three- and four-sided shapes
- Identify properties of 3D shapes
- Construct and interpret plans and elevations of 3D shapes
- Sketch nets of solids

</td><td>

SPECIFICATION LINKS

- G12, G13

</td></tr>
</table>

STARTER ACTIVITY

- **What am I?; 5 minutes; page 180**
  Tell the student that you are thinking of one of the 2D shapes on the page. The student may ask you questions about this shape but the only answers you can give are 'yes' or 'no'. They can ask up to 10 questions to correctly identify the shape. Repeat but with the student choosing a shape and you asking the questions.

MAIN ACTIVITIES

- **3D shapes; 25 minutes; page 181**
  For each 3D shape, ask the student to work out:
  a) the number of vertices   b) the number of edges   c) the number of faces.
  Establish that a prism is a 3D shape with the same cross-section all along its length. Invite the student to identify prisms in the room.
- **Construct; 15 minutes; page 182**
  Multilink cubes work well for this activity; alternatively, you could use dice stuck together with sticky tack.

PLENARY ACTIVITY

- **Name a shape; 5 minutes**
  Ask the student to name a shape that:
  a) has no vertices (sphere)   b) has 8 vertices (cube/cuboid)   c) has 1 edge (cone)
  d) has 6 edges (triangular-based prism)   e) has 2 faces (cone)   f) has 1 face (sphere)
  g) has 6 faces (cube/cuboid)   h) has 1 vertex (cone)

HOMEWORK ACTIVITY

- **Nets of a cube; 30 minutes; page 183**
  You may need to reinforce to the student that reflections and rotations of a net do not count as a different net.

SUPPORT IDEA

- **3D shapes, Construct** Provide the student with an object in each of the 3D shapes listed to help them count its vertices, edges and faces. You could ask them to find an object for each shape around the room or to think of real-life examples.

EXTENSION IDEAS

- **3D shapes** Challenge the student to work out how many planes of symmetry each shape has (cube = 9, cuboid = 3, cylinder = infinite, square-based pyramid = 4, triangular-based pyramid = 6, cone = infinite, sphere = infinite). You could also ask them to sketch a net for each shape.
- **Construct** Create more complex 3D shapes and ask the student to sketch the plan and front and side elevations.

PROGRESS AND OBSERVATIONS

STARTER ACTIVITY: WHAT AM I?

TIMING: 5 MINS

LEARNING OBJECTIVES
- Classify three- and four-sided shapes

EQUIPMENT
none

1. Guess which of these shapes your tutor is thinking of by asking up to 10 yes or no questions.

square	parallelogram	rhombus
trapezium	kite	rectangle
isosceles triangle	equilateral triangle	right-angled triangle
	scalene triangle	

MAIN ACTIVITY: 3D SHAPES

TIMING: 25 MINS

LEARNING OBJECTIVES

- Identify properties of 3D shapes

EQUIPMENT

1. **How many vertices, edges and faces does each of these shapes have?**
   Write your answers around the shapes.

cube 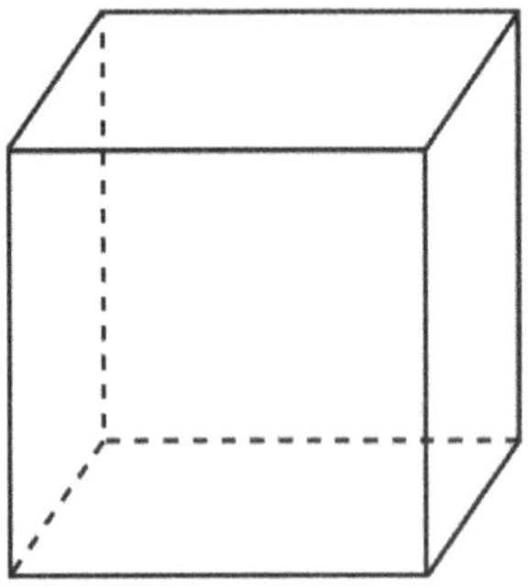	cuboid 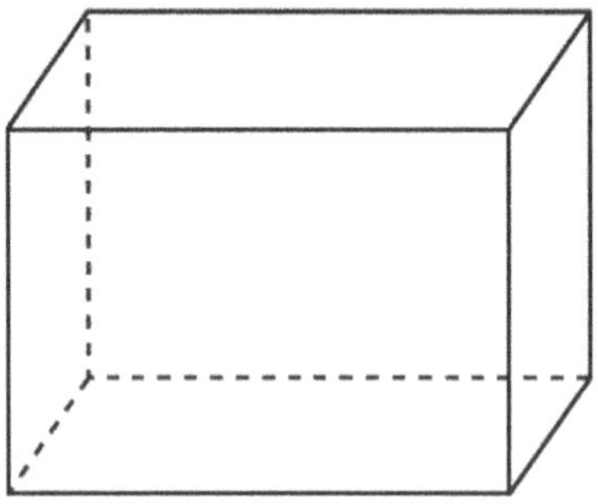	cylinder
square-based pyramid 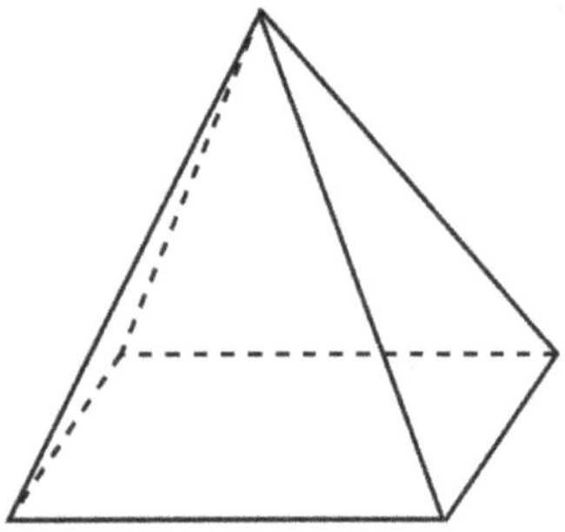	triangular-based pyramid 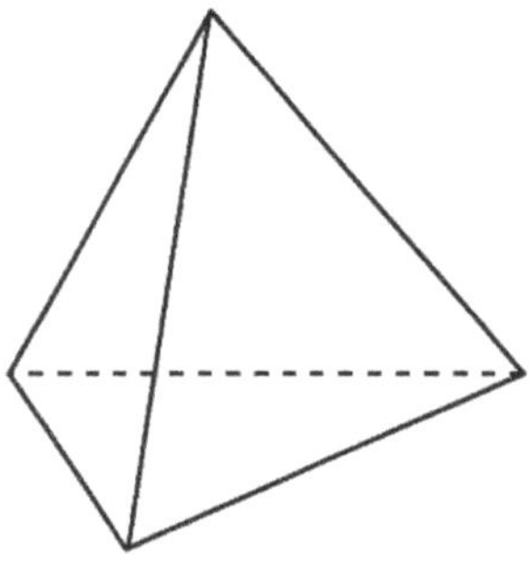	cone 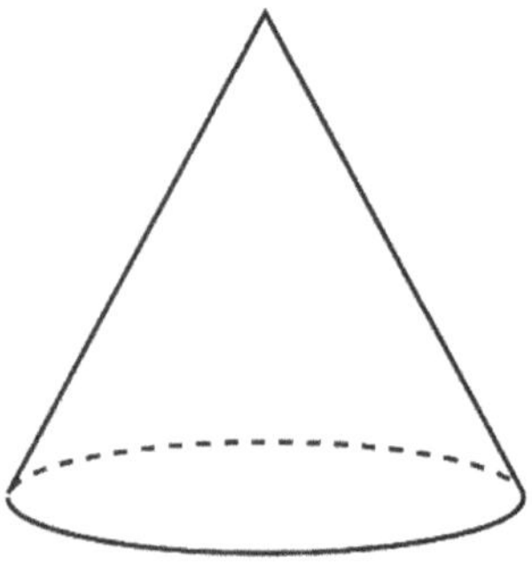
sphere 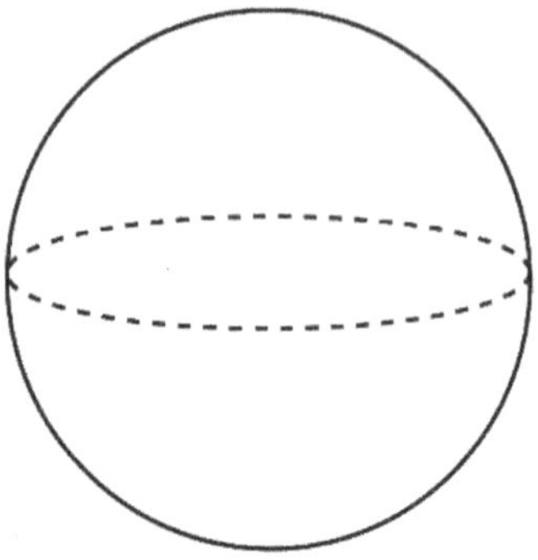	prisms 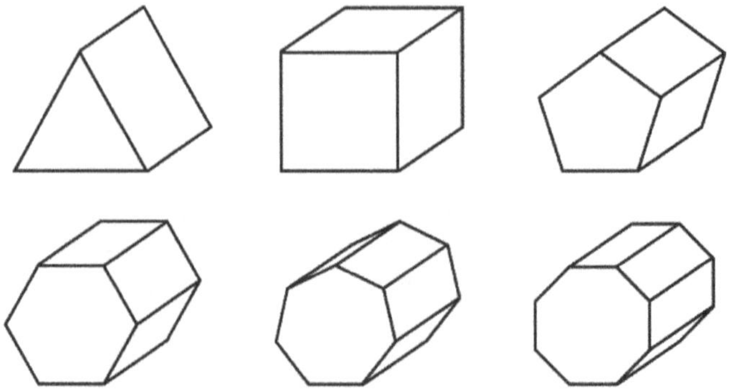	

MAIN ACTIVITY: CONSTRUCT

TIMING: 15 MINS

LEARNING OBJECTIVES

• Construct and interpret plans and elevations of 3D shapes

EQUIPMENT

• cubes

The **plan** of a shape is the view from above – think 'bird's eye view'.

The **front** and **side elevations** are the views from the front and side.

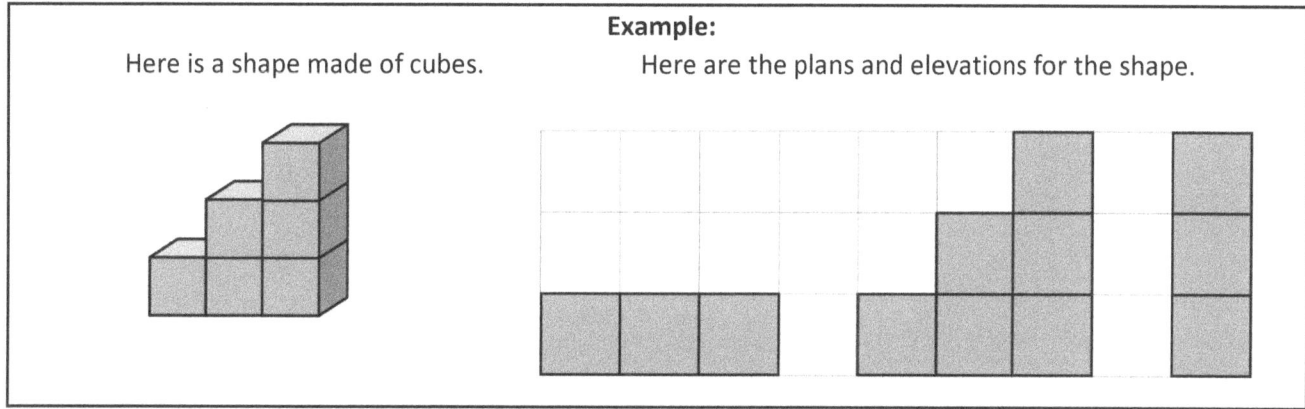

Example:

Here is a shape made of cubes.   Here are the plans and elevations for the shape.

1. Make or draw your own 3D shape made of cubes. Then sketch its plan, front and side elevation.

2. Make or sketch the 3D shape that has this plan, front elevation and side elevation.

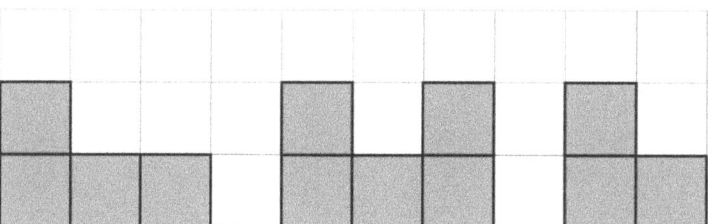

HOMEWORK ACTIVITY: NETS OF A CUBE

TIMING: 30 MINS

LEARNING OBJECTIVES

- Sketch nets of solids

EQUIPMENT

There are many different nets that fold to make a cube.

Here is one possible net.

1. **Draw as many different nets that fold to make a cube as you can.**

   Remember that rotations and reflections do not count as different nets.

MATHS
— FOUNDATION —

28 ANSWERS

STARTER ACTIVITY: WHAT AM I?

1. Check student names shapes correctly and can answer questions about their shape.

MAIN ACTIVITY: 3D SHAPES

1.

cube	cuboid	cylinder
8 vertices, 12 edges, 6 faces	8 vertices, 12 edges, 6 faces	0 vertices, 2 edges, 3 faces
square-based pyramid	**triangular-based pyramid**	**cone**
5 vertices, 8 edges, 5 faces	4 vertices, 6 edges, 4 faces	1 vertex, 1 edge, 2 faces
sphere	**prisms**	
0 vertices, 0 edges, 1 face	triangular: 6 vertices, 9 edges, 5 faces cuboid: 8 vertices, 12 edges, 6 faces pentagonal: 10 vertices, 15 edges, 7 faces hexagonal: 12 vertices, 18 edges, 8 faces heptagonal: 14 vertices, 21 edges, 9 faces octagonal: 16 vertices, 24 edges, 10 faces	

MAIN ACTIVITY: CONSTRUCT

1. Check student's answer.

2.

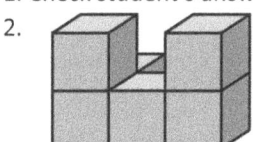

HOMEWORK ACTIVITY: NETS OF A CUBE

1. There are 11 possible nets for a cube. Here are the other ten:

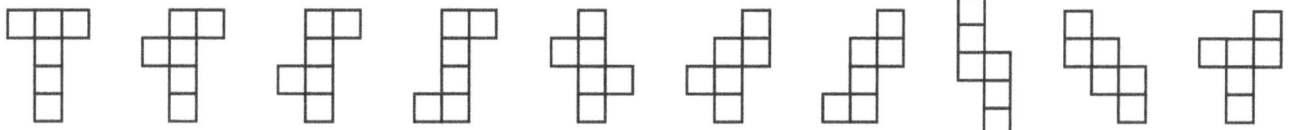

GLOSSARY

Plan view
The view of an object from directly above

Side/front elevation
The view of an object from the side/front

Net
A flat pattern than can be folded up to make a 3D shape

29 GEOMETRY AND MEASURES: AREA AND VOLUME

LEARNING OBJECTIVES

- Recognise units of measurement used for length, area and volume
- Know and apply formulae to calculate area of triangles, circles, parallelograms and trapezia
- Calculate the surface area of cubes, cuboids and prisms (including cylinders)
- Know and apply formulae to calculate the volume of cubes, cuboids and prisms (including cylinders)

SPECIFICATION LINKS

- G16, G17

STARTER ACTIVITY

- **Units of measure; 5 minutes; page 186**
  If necessary, discuss how to easily identify which units are used to measure length, area and volume.

MAIN ACTIVITIES

- **Calculating area; 25 minutes; page 187**
  Work through the table, completing the formula for the area of each of the shapes and marking the dimensions the formulae refer to on the relevant diagrams. Students are no longer given any of these formulae in the exam so they must learn them. Move on to calculating compound and surface areas. Remind the student of the formulae for finding the surface area of pyramids, cones and spheres.

- **Calculating volume; 15 minutes; page 188**
  Discuss the formula for calculating the volume of any prism. Look at the worked example, calculating the volume of the cylinder; explain that the exact cross-sectional area written in terms of π and the solution is not rounded until the final stage. Remind the student of the formulae for working out the volumes of pyramids, cones and spheres.

PLENARY ACTIVITY

- **Pop quiz; 5 minutes**
  Test the student on their memory of the formulae for areas, surface areas and volumes. Name a shape and property and ask the student to identify the formula, or give a formula and ask the student to say what shape and property it is for.

HOMEWORK ACTIVITY

- **Spider diagram; 60 minutes; page 189**
  Full instructions are given on the activity sheet.

SUPPORT IDEA

- **Calculating area** When calculating surface area, encourage the student to sketch a net of each shape before attempting to find the surface area. You could provide the student with actual 3D shapes to help them visualise the shapes of the different faces they need to calculate the area of. When finding the surface area of a cylinder, show that the curved side is a rectangle by peeling the label from a tin of soup or similar.

EXTENSION IDEAS

- **Calculating area** Challenge the student to find the surface area of a triangular prism whose cross-section is an equilateral triangle with sides of length 10 cm. They will need to use Pythagoras' theorem to find the height of the triangle before finding the surface area.
- **Calculating volume** Extend finding the volume of the shapes in cm^3 to working out how many litres each shape will hold.

PROGRESS AND OBSERVATIONS

STARTER ACTIVITY: UNITS OF MEASURE

TIMING: 5 MINS

LEARNING OBJECTIVES

- Recognise units of measurement used for length, area and volume

EQUIPMENT

1. Each of these units is a measure of length, area or volume. Sort them into the correct columns of the table.

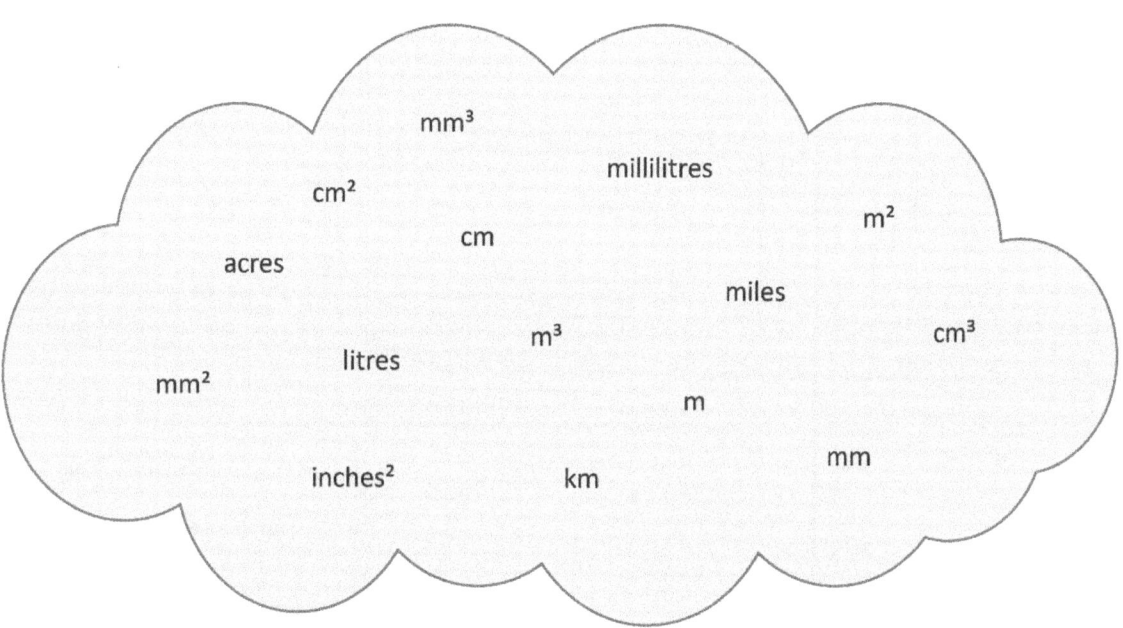

mm³
millilitres
cm²
m²
cm
acres
miles
m³
cm³
litres
mm²
m
inches²
km
mm

Length	Area	Volume

MATHS
— FOUNDATION —

MAIN ACTIVITY: CALCULATING AREA	TIMING: 25 MINS

LEARNING OBJECTIVES
- Know and apply formulae to calculate area of triangles, circles, parallelograms and trapezia
- Calculate the surface area of cubes, cuboids and prisms (including cylinders)

EQUIPMENT
- scissors

1. Write down the formula for calculating the area of each of these shapes. Mark the dimensions you use onto the diagrams. Cut out each card and fold back the formula. Use the cards to test yourself.

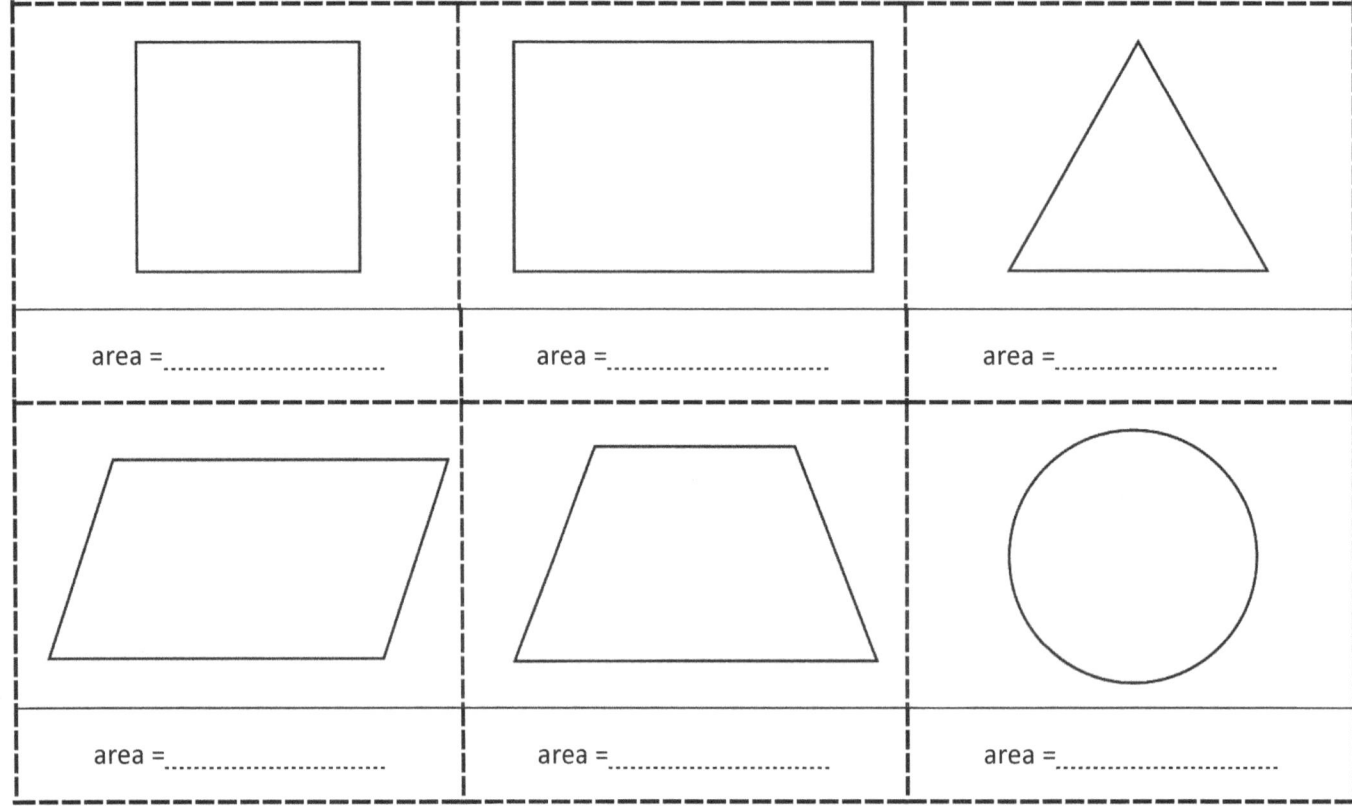

area =...............................   area =...............................   area =...............................

area =...............................   area =...............................   area =...............................

To work out the surface area of a 3D shape, try sketching the net. This will help you see how many faces there are and what shape each face is.

2. Work out the area or surface area of each of these shapes. Explain your method to your tutor.

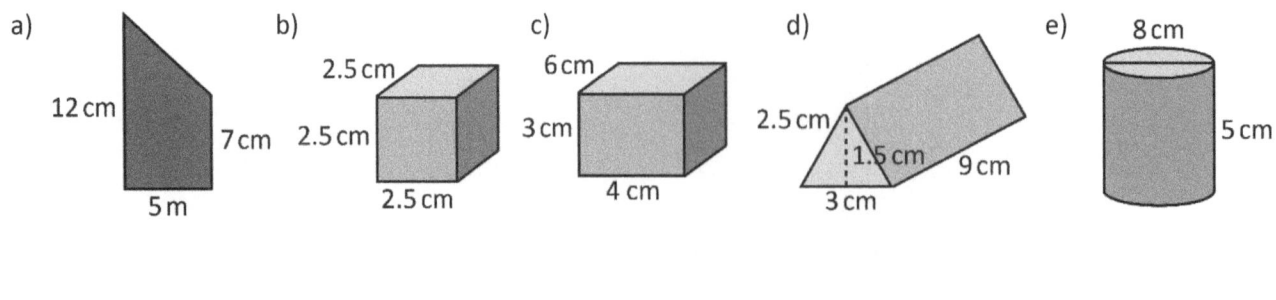

a) 12 cm 7 cm 5 m

b) 2.5 cm 2.5 cm 2.5 cm

c) 6 cm 3 cm 4 cm

d) 2.5 cm 1.5 cm 9 cm 3 cm

e) 8 cm 5 cm

....

MATHS
— FOUNDATION —

MAIN ACTIVITY: CALCULATING VOLUME	TIMING: **15** MINS

LEARNING OBJECTIVES

- Know and apply formulae to calculate the volume of cubes, cuboids and prisms (including cylinders)

EQUIPMENT

- calculator

volume of a prism = cross-sectional area × length

Example:

Calculate the volume of this cylinder to the nearest cm³.

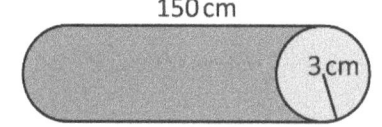

cross-sectional area = π × 3²   volume = 150 × 9π
                     = 9π   = 4241 cm³

---

1. Eric says *to work out the volume of a cuboid, you multiply the length by the width by the height.* Eric is correct. Explain why.

----

2. A cuboid has a volume of 280 cm³. Its cross-sectional area is 56 cm². Work out the depth of the cuboid.

----

3. How could you calculate the volume of these shapes?

a)

----

b)

----

c)

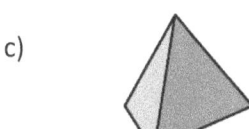

HOMEWORK ACTIVITY: SPIDER DIAGRAM

TIMING: 60 MINS

LEARNING OBJECTIVES

- Know and apply formulae to calculate area of triangles, circles, parallelograms and trapezia
- Know and apply the formula to calculate the area of a circle
- Calculate the surface area of cubes, cuboids and prisms (including cylinders)
- Know and apply formulae to calculate the volume of cubes, cuboids and prisms

EQUIPMENT

- large piece of paper

 1. Draw a spider diagram to display all the information you have learnt about calculating area and volume.

Ensure you include:

- how to calculate the area of squares, rectangles, circles, triangles, parallelograms and trapezia

- how to calculate the volume of cubes, cuboids, cylinders and prisms.

- how to calculate the surface area of cubes, cuboids, cylinders and triangular prisms

- which unit should be used for each measurement.

You should try to show where concepts link to one another. You may wish to include some examples.

29 ANSWERS

STARTER ACTIVITY: UNITS OF MEASURE

1.

Length	Area	Volume
cm	m^2	mm^3
miles	cm^2	millilitres
m	acres	cm^3
km	mm^2	m^3
mm	inches^2	litres

MAIN ACTIVITY: CALCULATING AREA

1. Check the student has labelled the correct dimensions on their diagrams.

square: area = b^2	rectangle: area = bh	triangle: area = $\frac{1}{2}bh$
parallelogram: area = bh	trapezium: area = $\frac{1}{2}(a + b)h$	circle: area = πr^2

2. a) 47.5 cm^2 b) 37.5 cm^2 c) 108 cm^2 d) 76.5 cm^2 e) 226.2 cm^2

MAIN ACTIVITY: CALCULATING VOLUME

1. The cross-sectional area is $b \times h$, and this is multiplied by the length.
2. 5 cm

3. a) $b \times b \times b$ b) $\pi r^2 h$ c) $\frac{1}{3} \times$ area of base $\times$ height

HOMEWORK ACTIVITY: SPIDER DIAGRAM

1. Student's own work

GLOSSARY

Cross-sectional area
The shape made when a 3D shape is cut through by a plane

30 GEOMETRY AND MEASURES: INTRODUCTION TO VECTORS

LEARNING OBJECTIVES

- Describe translations as 2D vectors
- Know the equations of simple straight line graphs parallel to the
  x- and *y*-axes
- Identify two column vectors that are parallel
- Identify, describe and construct congruent and similar shapes,
  including on coordinate axes, by considering rotation, reflection
  and translation

SPECIFICATION LINKS

- G7, G24, A9

STARTER ACTIVITY

- **Graphs parallel to the axis; 5 minutes; page 192**
  Full instructions are given on the activity sheet.

MAIN ACTIVITIES

- **Translations on a coordinate grid; 15 minutes; page 193**
  Remind the student of Lesson 26 when they covered translations. Recap how translations can be described using a
  column vector and illustrate this notation. Look at the diagram on the activity sheet. Ask the student to choose two
  shapes and describe in words the movement required to move one shape to the other (e.g. *how would you move
  shape A to shape B?*). Ask them to write this as a column vector. Repeat for other pairs of shapes. Then work through
  the questions on the activity sheet.
- **Vectors; 25 minutes; page 194**
  Show the student standard vector notation ($\overrightarrow{AB}$ or **a**) and challenge them to write the column vector that describes the
  movement between different pairs of points. Establish which vectors are equivalent ($\overrightarrow{BA}$ and $\overrightarrow{DC}$, $\overrightarrow{CE}$ and $\overrightarrow{BF}$, $\overrightarrow{HG}$ and $\overrightarrow{IC}$),
  and extend this to vectors $\overrightarrow{AB}$ and $\overrightarrow{CD}$, $\overrightarrow{EC}$ and $\overrightarrow{FB}$ and $\overrightarrow{GH}$ and $\overrightarrow{CI}$ therefore being equivalent. Agree that equivalent vectors
  are parallel, and show that you can see this by joining the points with straight lines. Show that parallel vectors may not be
  exactly the same length. Then ask the student to work through the questions on the activity sheet.

PLENARY ACTIVITY

- **Moving through a vector; 5 minutes**
  Place your student in the middle of the room and explain that this is the origin. Show them a column vector through
  which they must move and ask them to use one step to represent one unit. Move your student around the room. You
  could then challenge them to do the same to you, fixing a target end point.

HOMEWORK ACTIVITY

- **Transformations; 30 minutes; page 195**
  Full instructions are given on the activity sheet.

SUPPORT IDEA

- **Translations on a coordinate grid** Draw a horizontal arrow and a vertical arrow on the grid to show the movement of one
  vertex of the shape. Count the squares and ensure the student understands how this relates to the column vector.

EXTENSION IDEA

- **Vectors** Extend this activity by asking the student to show how they could express any vector parallel to a given vector
  (e.g. any vector $\begin{pmatrix} 3a \\ 2a \end{pmatrix}$ is parallel to $\begin{pmatrix} 3 \\ 2 \end{pmatrix}$).

PROGRESS AND OBSERVATIONS

STARTER ACTIVITY: GRAPHS PARALLEL TO THE AXIS   **TIMING: 5 MINS**

LEARNING OBJECTIVES

- Know the equations of simple straight line graphs parallel to the x- and y-axes

EQUIPMENT

- ruler

1. **Draw these lines on the coordinate grid below.**

   a) $x = -3$ b) $x = 4$ c) $y = -8$ d) $y = 5$

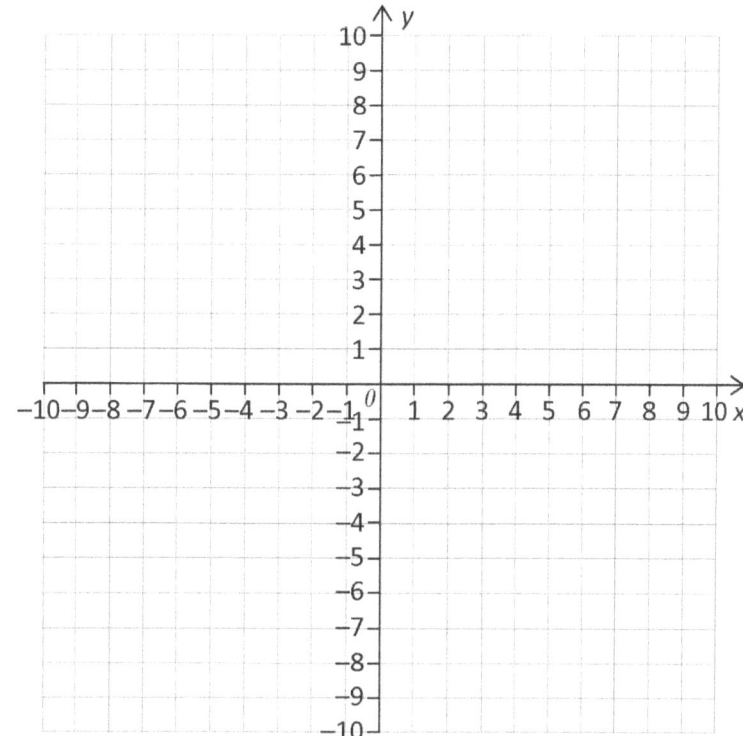

2. **Work out the area of the rectangle enclosed by these four lines.**

----

3. **Draw the graphs of $y = x$ and $y = -x$ on the same coordinate grid.**

MATHS
— FOUNDATION —

MAIN ACTIVITY: TRANSLATIONS ON A COORDINATE GRID   **TIMING: 15 MINS**

LEARNING OBJECTIVES

- Describe translations as 2D vectors

EQUIPMENT

- ruler

A vector has both magnitude and direction. It tells you how far and which way.

1. Look at the shapes on this coordinate grid.

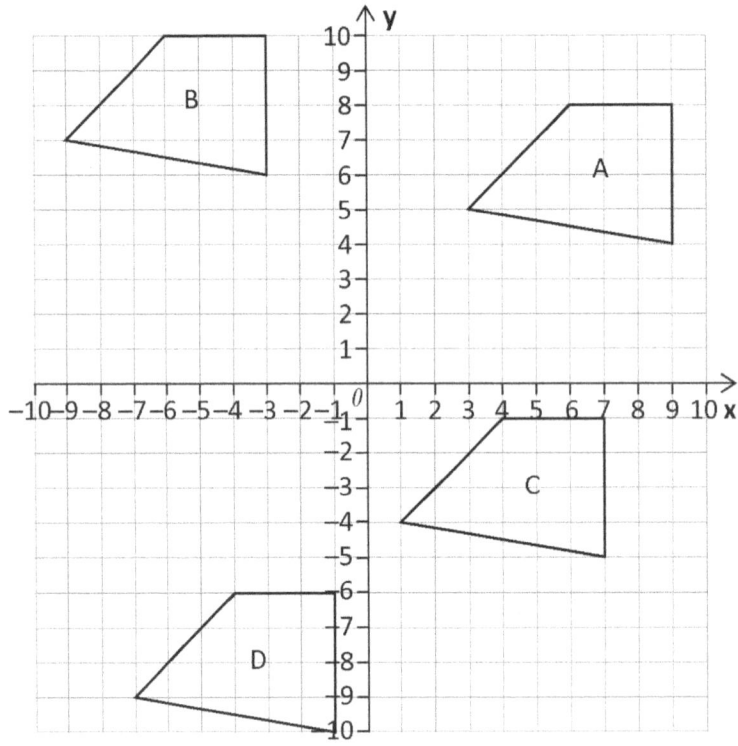

a) Translate shape A through vector $\begin{pmatrix} -6 \\ -4 \end{pmatrix}$. Label this image E.

b) How can you tell that the shapes A to E are all congruent to one another?

----

c) Join the corresponding corners of shapes A and B with straight lines. What do you notice about these lines? Explain why this is.

MAIN ACTIVITY: VECTORS

TIMING: 25 MINS

LEARNING OBJECTIVES

- Describe translations as 2D vectors
- Identify two column vectors that are parallel

EQUIPMENT

- ruler

1. Points A to I have been plotted on the coordinate grid below.

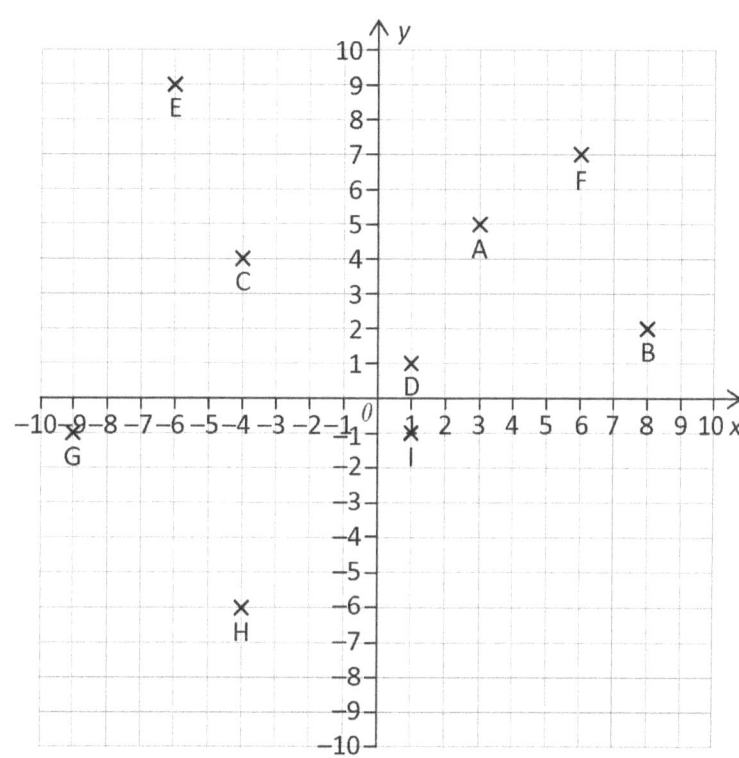

a) $\overrightarrow{BJ} = \begin{pmatrix} -1 \\ -2 \end{pmatrix}$. Mark point J on the diagram.

b) Write $\overrightarrow{JB}$ as a column vector. ....

c) Which vector on the diagram is parallel to vector $\overrightarrow{BJ}$? How could you work this out given just the column vector for each possible pair of points?

....

d) Write down three column vectors that would be parallel to $\overrightarrow{AF}$.

....

e) If you knew vector $\overrightarrow{XY}$, could you easily write down vector $\overrightarrow{YX}$?

....

MATHS
— FOUNDATION —

HOMEWORK ACTIVITY: TRANSFORMATIONS   **TIMING: 30 MINS**

LEARNING OBJECTIVES

- Describe translations as 2D vectors
- Identify, describe and construct congruent and similar shapes, including on coordinate axes, by considering rotation, reflection and translation

EQUIPMENT

- ruler

1. Shape A has been drawn on the coordinate grid below.

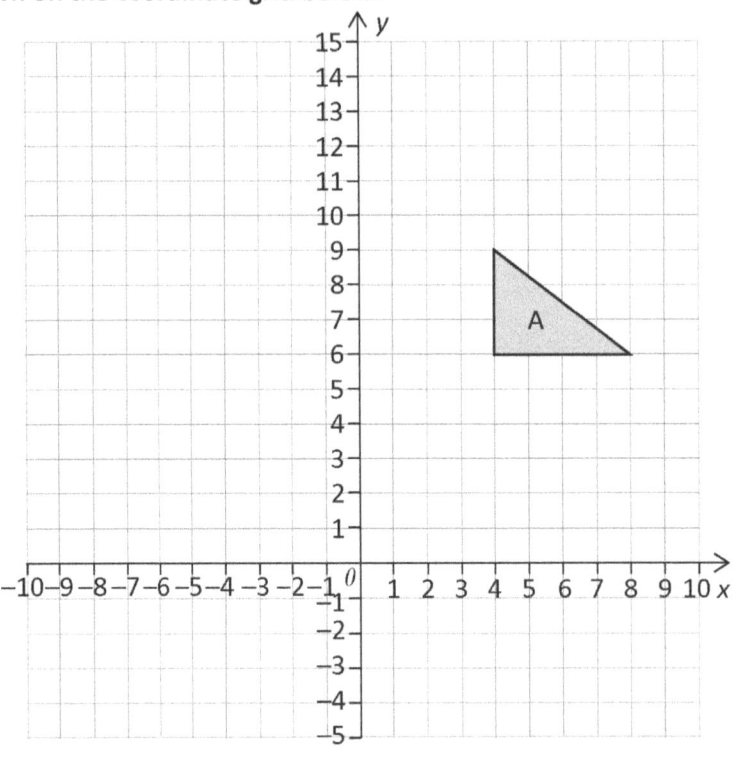

a) Translate shape A through $\begin{pmatrix} -5 \\ 2 \end{pmatrix}$. Label the shape B.

b) Reflect B in the line x = −2. Label the shape C.

c) Rotate C through 90° about the point (0, 8). Label the shape D.

d) Reflect D in the line y = 8. Label the shape E.

e) Rotate E 270° about the point (−2, 0). Label the shape F.

f) Describe the single transformation that will move shape F to shape A.

----

g) What can you say about all the triangles on the diagram?

----

h) Write down three vectors parallel to the vector that moved A to B. -----------------------------

STARTER ACTIVITY: GRAPHS PARALLEL TO THE AXIS

1. Check student's coordinate grid.
2. 91 units^2
3. Check student's coordinate grid.

MAIN ACTIVITY: TRANSLATIONS ON A COORDINATE GRID

1. a) Shape E drawn with vertices at (−3, 1), (3, 0), (3, 4) and (0, 4)
b) The shape and size are unchanged by translation.
c) The lines are parallel since each corner has moved through the same vector.

MAIN ACTIVITY: VECTORS

1. a) Point J plotted at (7, 0) b) $\begin{pmatrix} 1 \\ 2 \end{pmatrix}$ c) $\overrightarrow{AD}$. $\overrightarrow{AD} = \begin{pmatrix} -2 \\ -4 \end{pmatrix}$, which is a multiple of $\begin{pmatrix} -1 \\ -2 \end{pmatrix}$.

d) Any three vectors of the form $\begin{pmatrix} 3a \\ 2a \end{pmatrix}$ e) Change the signs of both parts of the vector.

HOMEWORK ACTIVITY: TRANSFORMATIONS

1. a)–e)

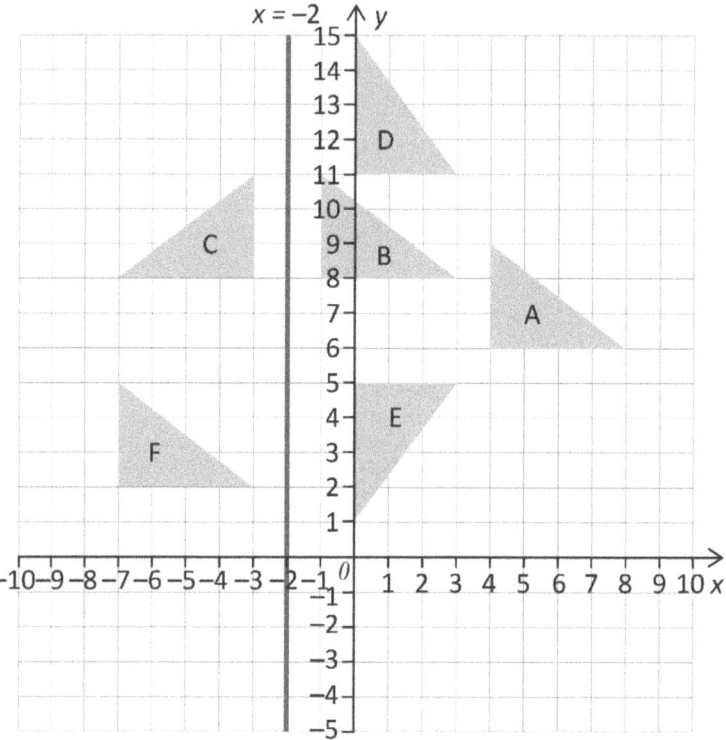

f) Translate F through vector $\begin{pmatrix} 11 \\ 4 \end{pmatrix}$.

g) They are all congruent.

h) Any vectors of the form $\begin{pmatrix} -5a \\ 2a \end{pmatrix}$

GLOSSARY

Vector
A vector describes displacement. It has both magnitude (or size) and direction.

31 GEOMETRY AND MEASURES: CALCULATING WITH VECTORS

LEARNING OBJECTIVES

- Add and subtract vectors
- Multiply vectors by a scalar
- Represent vectors diagrammatically and using column notation

SPECIFICATION LINKS

- G25, N2, N3

STARTER ACTIVITY

- **Working with negative numbers; 5 minutes; page 198**
  Full instructions are given on the activity sheet. Cut out the cards before the lesson.

MAIN ACTIVITIES

- **Drawing vectors; 20 minutes; page 199**
  Remind the student of the definition of a vector and how to represent this information as a column vector. Look at the vectors **x** and **y** on the activity sheet and establish that they are parallel but not equal. Then invite the student to represent some of the vectors **a** to **f** on the diagram and discuss whether any are equal or parallel (none are).
  Ask the student to draw vector **a** followed by vector **b**, then to work out how the vector **a** + **b** could be represented as a column vector. Repeat for **a** – **b**. Repeat for other pairs of vectors; discuss how we can add or subtract the vectors without drawing the line segments (by adding or subtracting the parts of the vectors).
  Ask the student to draw two lots of vector **a** and explain that this could be written as **2a** and can be found by multiplying each part of the vector by 2. Repeat for other multiples of different vectors.

- **Calculating with vectors; 20 minutes; page 200**
  Work through the worked example with the student, explaining that calculating with vectors is very similar to working with numbers. Then ask the student to complete the questions.

PLENARY ACTIVITY

- **Mind map; 5 minutes**
  To support the homework, spend five minutes making a mind map of all the things the student knows about vectors.

HOMEWORK ACTIVITY

- **Teach it!; 45 minutes; page 201**
  Full instructions are given on the activity sheet.

SUPPORT IDEA

- **Calculating with vectors** If the student found the starter challenging, spend some time reinforcing how to add, subtract, multiply and divide with negative numbers before starting this activity. You may also wish to draw the vectors on squared paper to help the student visualise the addition/subtraction/multiplication of the vectors.

EXTENSION IDEA

- **Calculating with vectors** Challenge the student to draw shapes using a combination of the vectors provided, drawing their answers and writing them in vector form.

PROGRESS AND OBSERVATIONS

STARTER ACTIVITY: WORKING WITH NEGATIVE NUMBERS   TIMING: 5 MINS

LEARNING OBJECTIVES

- Add, subtract, multiply and divide positive and negative integers

EQUIPMENT

- scissors

1. Cut out all three sets of cards below. Use three of the number cards and two of the operators to make:

a) the largest number possible

b) the smallest number possible

c) the number closest to zero.

Use three number cards:

0	1	2	3	4	5	6	7	8	9
	−1	−2	−3	−4	−5	−6	−7	−8	−9

Use two operators:

+	−	×	÷

Use brackets if you need them:

(	)

MATHS
— FOUNDATION —

AQA

MAIN ACTIVITY: DRAWING VECTORS

TIMING: 20 MINS

LEARNING OBJECTIVES

- Add and subtract vectors
- Multiply vectors by a scalar
- Represent vectors diagrammatically and using column notation

EQUIPMENT

- ruler

A vector can be shown as a line segment. The vectors $\mathbf{x} = \begin{pmatrix} 2 \\ -4 \end{pmatrix}$ and $\mathbf{y} = \begin{pmatrix} 3 \\ -6 \end{pmatrix}$ are shown on the diagram below.

Equal vectors have the same **magnitude and direction**, parallel vectors have the same **direction**.

As you can see on the graph, **x** and **y** are parallel vectors.

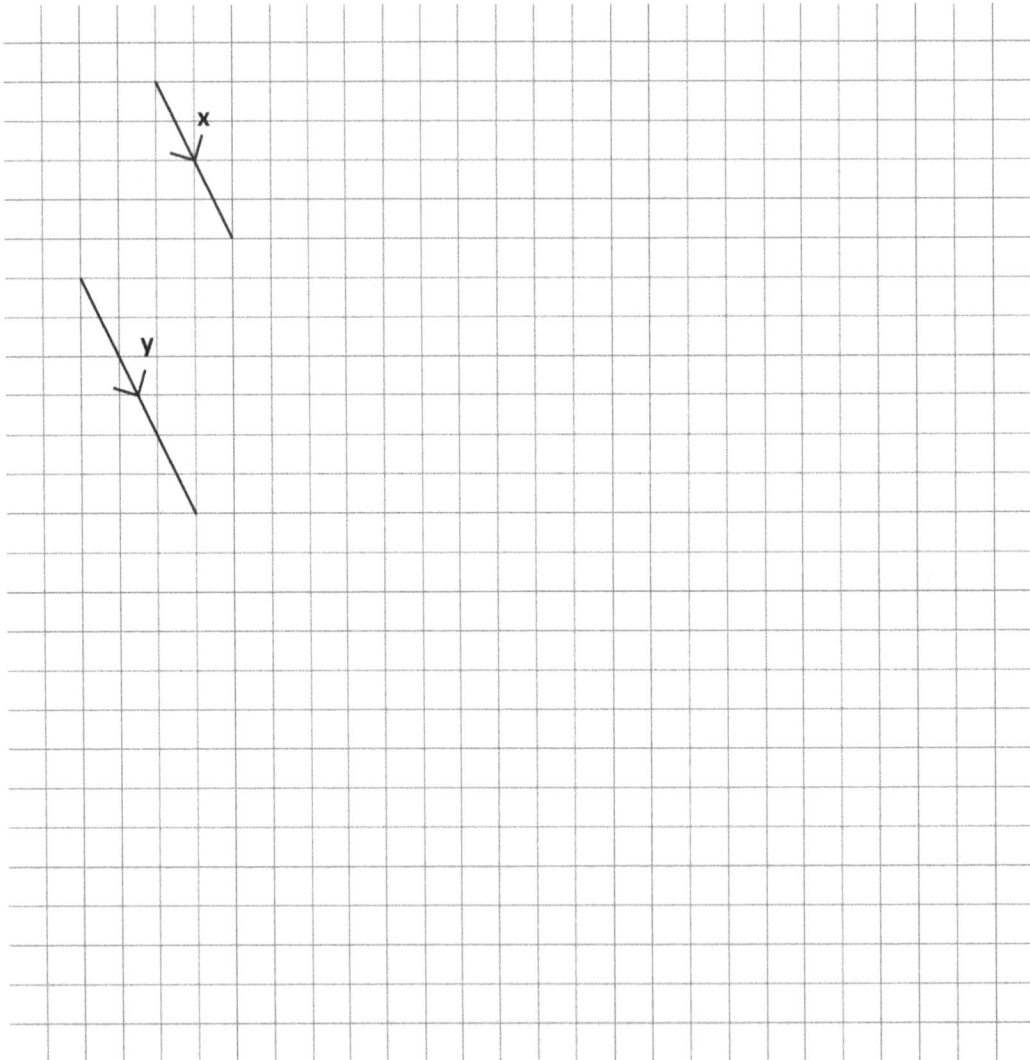

 1. On the graph, draw a line segment to represent each of these vectors.

$$\mathbf{a} = \begin{pmatrix} 3 \\ 5 \end{pmatrix} \qquad \mathbf{b} = \begin{pmatrix} 1 \\ -4 \end{pmatrix} \qquad \mathbf{c} = \begin{pmatrix} -5 \\ -2 \end{pmatrix} \qquad \mathbf{d} = \begin{pmatrix} 0 \\ 6 \end{pmatrix} \qquad \mathbf{e} = \begin{pmatrix} -7 \\ 0 \end{pmatrix} \qquad \mathbf{f} = \begin{pmatrix} -3 \\ 2 \end{pmatrix}$$

MAIN ACTIVITY: CALCULATING WITH VECTORS   TIMING: 20 MINS

LEARNING OBJECTIVES

- Add and subtract vectors
- Multiply vectors by a scalar
- Represent vectors diagrammatically and using column notation

EQUIPMENT

Example: $m = \begin{pmatrix} 4 \\ 2 \end{pmatrix}$ $n = \begin{pmatrix} -3 \\ 7 \end{pmatrix}$ $q = \begin{pmatrix} 0 \\ -4 \end{pmatrix}$

Work out: a) $m + n$ b) $n - q$ c) $3n$ d) $\frac{1}{2}m + q$

a) $\begin{pmatrix} 4 \\ 2 \end{pmatrix} + \begin{pmatrix} -3 \\ 7 \end{pmatrix} = \begin{pmatrix} 4 + -3 \\ 2 + 7 \end{pmatrix} = \begin{pmatrix} 1 \\ 9 \end{pmatrix}$

b) $\begin{pmatrix} -3 \\ 7 \end{pmatrix} - \begin{pmatrix} 0 \\ -4 \end{pmatrix} = \begin{pmatrix} -3 - 0 \\ 7 - -4 \end{pmatrix} = \begin{pmatrix} -3 \\ 11 \end{pmatrix}$

c) $3 \times \begin{pmatrix} -3 \\ 7 \end{pmatrix} = \begin{pmatrix} 3 \times -3 \\ 3 \times 7 \end{pmatrix} = \begin{pmatrix} -9 \\ 21 \end{pmatrix}$

d) $\frac{1}{2} \times \begin{pmatrix} 4 \\ 2 \end{pmatrix} + \begin{pmatrix} 0 \\ -4 \end{pmatrix} = \begin{pmatrix} (4 \div 2) + 0 \\ (2 \div 2) + -4 \end{pmatrix} = \begin{pmatrix} 2 \\ -3 \end{pmatrix}$

Look at these vectors: $a = \begin{pmatrix} 3 \\ -5 \end{pmatrix}$ $b = \begin{pmatrix} 2 \\ 7 \end{pmatrix}$ $c = \begin{pmatrix} -2 \\ 8 \end{pmatrix}$ $d = \begin{pmatrix} 0 \\ -5 \end{pmatrix}$ $e = \begin{pmatrix} -1 \\ 1 \end{pmatrix}$ $f = \begin{pmatrix} -4 \\ -14 \end{pmatrix}$

1. Work out these vector calculations.

   a) $a + b$ ....

   b) $d - e$ ....

   c) $2c - 3d$ ....

   d) $3c$ ....

   e) $2a - e$ ....

   f) $\frac{1}{2}c$ ....

2. Work out these vectors.

   a) $-a$ ....

   b) $-b$ ....

   c) $-c$ ....

3. Give reasons for your answers to question 2.

----

4. Two of the vectors above are parallel. Find them. Give a reason for your answer.

HOMEWORK ACTIVITY: TEACH IT!

TIMING: 45 MINS

LEARNING OBJECTIVES

- Add and subtract vectors
- Multiply vectors by a scalar
- Represent vectors diagrammatically and using column notation

EQUIPMENT

- squared and plain paper
- ruler

 1. **Your task is to design a page of a textbook to teach a student all about vectors. Imagine the student has never seen vectors before.**

Ensure you include:

- the definition of a vector

- the different ways a vector can be represented and drawn

- how to add/subtract/multiply vectors

- worked examples

- an exercise with the answers.

31 ANSWERS

STARTER ACTIVITY: WORKING WITH NEGATIVE NUMBERS

1. a) $(8 - -9) \times 9 = 153$ b) $(-9 + -8) \times 9 = -153$ c) Any calculation multiplied by 0 to give 0

MAIN ACTIVITY: DRAWING VECTORS

1.

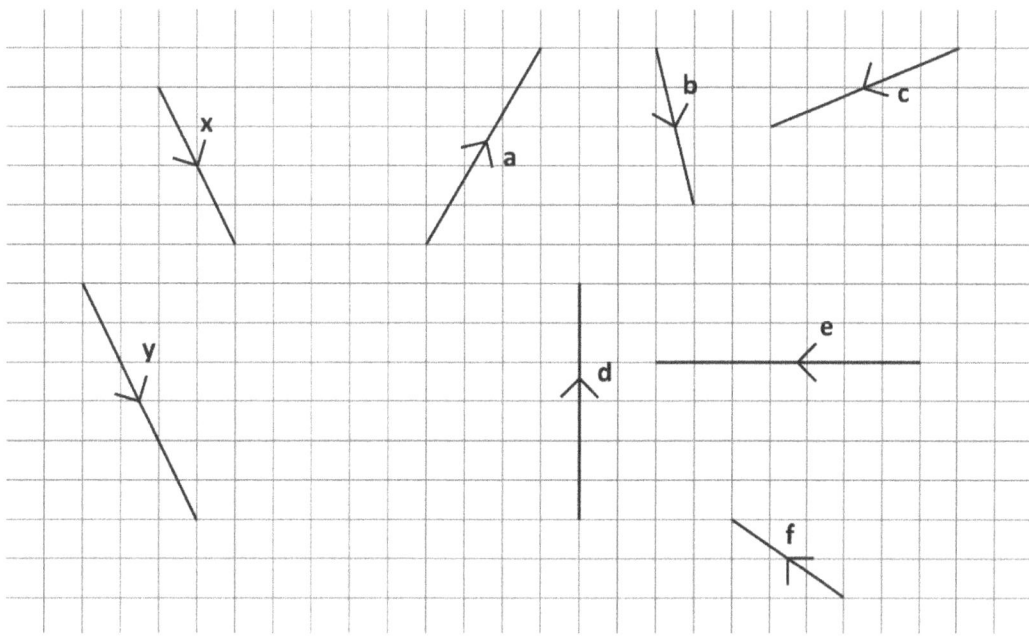

MAIN ACTIVITY: CALCULATING WITH VECTORS

1. a) $\begin{pmatrix} 5 \\ 2 \end{pmatrix}$ b) $\begin{pmatrix} 1 \\ -6 \end{pmatrix}$ c) $\begin{pmatrix} -4 \\ 31 \end{pmatrix}$ d) $\begin{pmatrix} -6 \\ 24 \end{pmatrix}$ e) $\begin{pmatrix} 7 \\ -11 \end{pmatrix}$ f) $\begin{pmatrix} -1 \\ 4 \end{pmatrix}$

2. a) $\begin{pmatrix} -3 \\ 5 \end{pmatrix}$ b) $\begin{pmatrix} -2 \\ -7 \end{pmatrix}$ c) $\begin{pmatrix} 2 \\ -8 \end{pmatrix}$

3. Change the sign of both parts of the vector.

4. **b** and **f** because they are multiples of one another.

HOMEWORK ACTIVITY: TEACH IT!

1. Student's own work

32 PROBABILITY: BASIC PROBABILITY AND VENN DIAGRAMS

LEARNING OBJECTIVES

- Order positive integers, decimals and fractions
- Record, describe and analyse the frequency of outcomes of probability experiments using tables and frequency trees
- Relate relative expected frequencies to theoretical probability, using appropriate language and the 0–1 probability scale
- Apply the property that the probabilities of an exhaustive set of outcomes sum to one
- Enumerate sets and combinations of sets systematically, using tables, grids and Venn diagrams

SPECIFICATION LINKS

- N1, P1, P2, P3, P4, P5, P6

STARTER ACTIVITY

- **Fractions, decimals and percentages; 5 minutes; page 204**
  Remind the student how to convert between fractions, decimals and percentages before working through the activity.

MAIN ACTIVITIES

- **Colouring in; 25 minutes; page 205**
  Explain to the student that each grid must be coloured in to satisfy the given statement. Encourage the student to recognise that there is often more than one way to do this.
- **Venn diagrams; 15 minutes; page 206**
  Discuss what a Venn diagram is and work through the information on the activity sheet, focusing on ensuring the student understands set notation. Explain that Venn diagrams can be used to help with probability calculations.

PLENARY ACTIVITY

- **Rolling a dice; 5 minutes**
  Ask the student to roll a dice 10 times and to tally the results. Use this tally to work out the experimental probability of rolling different values, and discuss whether this is equal to the theoretical probability. Roll another 20 times and compare experimental probability to theoretical probability again. Establish that the more times the dice is rolled, the closer the experimental probability should become to the theoretical probability.

HOMEWORK ACTIVITY

- **Exam-style questions; 15 minutes; page 207**
  Before setting the homework, remind the student of the differences between replacement and non-replacement probability questions. Ask them to think of some examples where the probability might change if something was chosen and then removed from the list of options.

SUPPORT IDEA

- **Venn diagrams** Draw a Venn diagram with two overlapping sectors labelled A and B. Ask the student to shade the areas that represent A, B, A∩B, A∪B, A' and B'.

EXTENSION IDEA

- **Colouring in** Challenge the student to find as many different ways as they can to satisfy the criteria for each different question. Which ones have an infinite number of options?

PROGRESS AND OBSERVATIONS

STARTER ACTIVITY: FRACTIONS, DECIMALS AND PERCENTAGES TIMING: 5 MIN

LEARNING OBJECTIVES	EQUIPMENT
• Order positive integers, decimals and fractions	none

1. Probabilities can be written as fractions, decimals or percentages. Write these values in ascending order.

0.3	$\frac{2}{5}$	35%	0.303	$\frac{16}{50}$	31%

....

MATHS
— FOUNDATION —

MAIN ACTIVITY: COLOURING IN	TIMING: 25 MINS

LEARNING OBJECTIVES

- Relate relative expected frequencies to theoretical probability, using appropriate language and the 0–1 probability scale
- Apply the property that the probabilities of an exhaustive set of outcomes sum to one

EQUIPMENT

- colouring pencils

1. **Colour each grid with the appropriate colours to satisfy the given statements or tables. Assume that squares are picked at random.**

a) The probability of picking a blue square is $\frac{3}{4}$.

b) The probability of not picking a green square is $\frac{3}{5}$.

c) The squares are all blue or red. The probability of picking a blue square is twice that of picking a red square.

d) The probability of picking each coloured square is:
blue: $\frac{1}{4}$, red: $\frac{1}{3}$, yellow: $\frac{1}{12}$

e)

Colour	blue	red	yellow
Tally	///// /	///// ///// ///// /	//

f)

Colour	blue	red	yellow	green
Probability	0.2	0.1	0.3	0.4

g) Alice picked a square 20 times. She picked a green square 4 times, a red square 14 times and a blue square 2 times.

h) Nav picked a square 100 times. The number of red squares picked was 20.

i)

Colour	blue	red	yellow
Probability	x	$3x$	0.2

MAIN ACTIVITY: VENN DIAGRAMS

TIMING: **15** MINS

LEARNING OBJECTIVES

- Enumerate sets and combinations of sets systematically, using Venn diagrams

EQUIPMENT

A set of values is a group of numbers. You can show the members of a set by listing them inside curly brackets.

If X is the set of even numbers smaller than 10 we can write this as: X = {2, 4, 6, 8}

If Y is the set of prime numbers smaller than 10 we can write this as: Y = {2, 3, 5, 7}

Learn these symbols:

Symbol	Meaning	Example
∈	belongs to	2 is in the set X so we can say 2 ∈ X
X′	not X	X′ is all the numbers being considered that are not in the set X X′ = {1, 3, 5, 7, 9}
∪	union (or)	X ∪ Y is all the elements in set X or in set Y X ∪ Y = {2, 3, 4, 5, 6, 7, 8}
∩	intersect (and)	X ∩ Y is all the elements in set X **and** in set Y X ∩ Y = {2}
ξ	the universal set (all elements being considered)	ξ = positive whole numbers smaller than 10

1. **Complete this Venn diagram for sets X and Y as described above.**

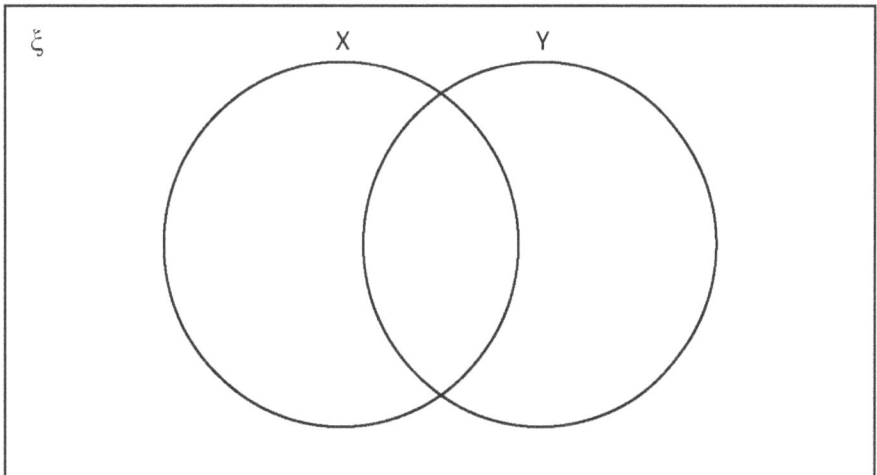

2. **A positive integer than 10 is chosen at random. Work out these probabilities.**

a) P(X) ....

b) P(X ∪ Y) ....

c) P(Y′) ....

d) P (X ∩ Y) ....

MATHS
— FOUNDATION —

| HOMEWORK ACTIVITY: EXAM-STYLE QUESTIONS | TIMING: 15 MINS |

LEARNING OBJECTIVES

- Record, describe and analyse the frequency of outcomes of probability experiments using tables and frequency trees
- Relate relative expected frequencies to theoretical probability, using appropriate language and the 0–1 probability scale
- Apply the property that the probabilities of an exhaustive set of outcomes sum to one

EQUIPMENT

 1. Here is a fair spinner. Henry spins it once.

a) Mark the probability that the spinner will land on an even number.

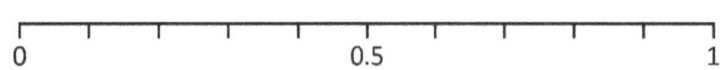

| | | | | | | | | | | |
0   0.5   1

b) Mark the probability that the spinner will land on a prime number.

| | | | | | | | | | | |
0   0.5   1

 2. In a bag of sweets, there are four different colours. There are 20 red sweets, 12 blue sweets, 7 green sweets and 1 yellow sweet. April takes a sweet at random.

a) Write down the probability that she takes a green sweet. ....

b) Write down the probability that she takes a yellow or blue sweet. ....

c) She takes a green sweet and eats it. She then takes another sweet. ....
   What is the probability that the second sweet is red?

 3. Here is a Venn diagram of sets A, B and ξ.

a) Write down the numbers that are in these sets.

   i)   A ∪ B   ....

   ii)   B'   ....

b) A number is chosen at random from the diagram. What is the probability that the number is in set A ∩ B?

....

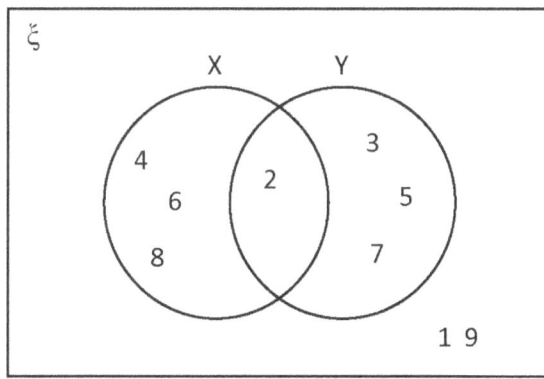

MATHS — FOUNDATION — AQA

STARTER ACTIVITY: FRACTIONS, DECIMALS AND PERCENTAGES

1. 0.3   0.303   31% $\frac{16}{50}$ 35% $\frac{2}{5}$

MAIN ACTIVITY: COLOURING IN

1.

a) 9 blue squares other squares any other colour	b) 8 green squares other squares any other colour	c) 8 blue squares 4 red squares
d) 3 blue squares 4 red squares 1 yellow square other squares any other colour	e) 6 blue squares 16 red squares 2 yellow squares	f) 6 blue squares 3 red squares 9 yellow squares 12 green squares
g) 2 squares green 7 squares red 1 square blue	h) 2 red squares other squares any other colour	i) 2 blue squares 6 red squares 2 yellow squares

MAIN ACTIVITY: VENN DIAGRAMS

1.

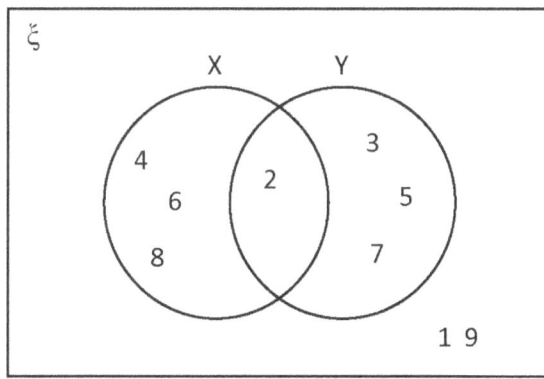

2. Allow equivalent fractions or percentages for all four parts.

a) $\frac{4}{9}$ b) $\frac{7}{9}$ c) $\frac{5}{9}$ d) $\frac{1}{9}$

HOMEWORK ACTIVITY: EXAM-STYLE QUESTIONS

1. a) mark at 0.5   b) mark at 0.4

2. a) $\frac{7}{40}$ b) $\frac{13}{40}$ c) $\frac{20}{39}$

3. a) i) 1, 2, 3, 4, 5, 7, 8, 9   ii) 0, 3, 4, 5, 6, 10   b) $\frac{2}{11}$

33 PROBABILITY: COMBINED PROBABILITY

LEARNING OBJECTIVES

- Apply systematic listing strategies
- Construct theoretical possibility spaces for single and combined experiments with equally likely outcomes and use these to calculate theoretical probabilities
- Calculate the probability of independent and dependent combined events, including using tree diagrams and other representations, and know the underlying assumptions

SPECIFICATION LINKS

- P7, P8, N5

STARTER ACTIVITY

- **How many bunnies?; 5 minutes; page 210**
  Encourage the student to list the outcomes systematically to ensure they include all possible outcomes. For the second part, guide the student towards multiplying the number of different sizes by the number of different chocolates.

MAIN ACTIVITIES

- **Two or more events; 20 minutes; page 211**
  Look at the frequency tree together, asking the student to explain where the numerical values come from. Extend to the two-way table and the associated probabilities.
- **Tree diagrams; 20 minutes; page 212**
  Explain that tree diagrams are similar to frequency trees, except that the probabilities are written on the branches. Model how a tree diagram is drawn, explaining that the number of 'branches' is equal to the number of possible outcomes. Explain the difference between dependent and independent events and ask the student to give some examples.
  Reinforce that the sum of the probabilities on each column should be 1.
  Work through the activity sheet, stressing the difference between the two situations (with or without replacement).

PLENARY ACTIVITY

- **Is it fair?; 5 minutes**
  Explain to the student that you are going to play a game with them. Produce two dice and say that you will roll the dice and add together the values. If the outcome is 6 or more you win, if it is less than 6 they win. Play a few rounds and ask them to decide if the game is fair or not. Ask them to try to explain why it is unfair.

HOMEWORK ACTIVITY

- **A fair game; 30 minutes; page 213**
  This activity depends on the plenary. Full instructions are given on the activity sheet.

SUPPORT IDEA

- **Tree diagrams** Model the problem using marbles or coloured pens or pencils to help the student visualise it. You could use coloured sweets and eat the first one.

EXTENSION IDEA

- **Two or more events** Ask the student to draw up their own two-way table for a game where two dice are rolled and their product is found. Then ask questions like *'what is the probability that you will score more than 10?'*

PROGRESS AND OBSERVATIONS

STARTER ACTIVITY: HOW MANY BUNNIES?   TIMING: 5 MINS

LEARNING OBJECTIVES

- Apply systematic listing strategies

EQUIPMENT

1. A factory produces chocolate bunnies in three sizes: small, medium and large.
   They use two different types of chocolate: dark chocolate and milk chocolate.

   How many different types of chocolate bunny are made? List them all.

   ....

   ....

   ....

2. The factory decides to introduce white chocolate in addition to milk and dark.
   How many different bunnies are there now? How can you work out the answer without listing all
   the possibilities?

   ....

   ....

   ....

MATHS
— FOUNDATION —

MAIN ACTIVITY: TWO OR MORE EVENTS

TIMING: 20 MINS

LEARNING OBJECTIVES

- Construct theoretical possibility spaces for single and combined experiments with equally likely outcomes and use these to calculate theoretical probabilities
- Calculate the probability of independent and dependent combined events

EQUIPMENT

A frequency tree shows how many times two or more events occur.

1. **A gardener grows 50 plants. 35 are strawberry plants and 15 are raspberry canes.**

   Of the strawberry plants, 20 grow fruit. Of the raspberry canes, 10 grow fruit. This information is displayed in this frequency tree.

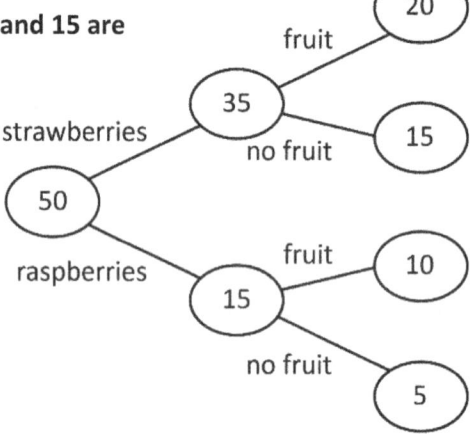

   a) Explain each of the values shown in the example frequency tree.

   ----

   b) This data can also be represented in a two-way table. Complete the table below.

	Strawberries	Raspberries	Total
Fruit			
No fruit			
Total			

2. **The gardener chooses a plant at random. What is the probability that it is:**

   a) a strawberry plant

   b) a fruiting raspberry

   c) a fruiting plant?

   ....

3. **The gardener chooses a fruiting plant. What is the probability that it is:**

   a) a raspberry   ....   b) a strawberry?   ....

MAIN ACTIVITY: TREE DIAGRAMS   TIMING: 20 MINS

LEARNING OBJECTIVES

- Construct theoretical possibility spaces for single and combined experiments with equally likely outcomes and use these to calculate theoretical probabilities
- Calculate the probability of independent and dependent combined events, including using tree diagrams and other representations, and know the underlying assumptions

EQUIPMENT

1. **A bag of marbles contains 3 red marbles and 8 green marbles. Adrian takes a marble at random, records the colour and puts it back in the bag. He then takes another marble.**

   a) Complete the tree diagram showing the probabilities and possible outcomes by writing the probability of each event as a fraction on each branch.

   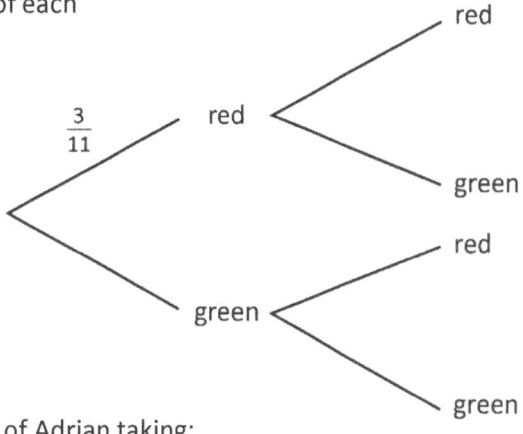

   b) Use your tree diagram to work out the probability of Adrian taking:

      i)   two green marbles   ....

      ii) one marble of each colour   ....

      iii) at least one green marble.   ....

   c) If Adrian had kept the first marble, would the tree be different? If so, explain how.

   d) Work out the answers to part b) for if Adrian had kept the marble.

      i)   two green marbles   ....

      ii) one marble of each colour   ....

      iii) at least one green marble   ....

MATHS
— FOUNDATION —

HOMEWORK ACTIVITY: A FAIR GAME

TIMING: 30 MINS

LEARNING OBJECTIVES	EQUIPMENT
• Calculate the probability of independent and dependent combined events	none

1. **At the end of the lesson you played a game where you rolled two dice and added the numbers to work out the score. If the score was 6 or more your tutor won, if it was less than 6 you won.**

   a) Complete the sample space diagram to show all the possible outcomes.

		Dice 1					
		1	2	3	4	5	6
Dice 2	1	2	3	4			
	2	3	4				
	3						
	4						
	5						
	6						

   b) Explain why the game is unfair.

   ....

   c) How could you change the rules to make this game fair?

   ....

   d) Another game involves multiplying together the values on two dice. Complete a sample space diagram for this game.

		Dice 1					
		1	2	3	4	5	6
Dice 2	1						
	2						
	3						
	4						
	5						
	6						

   e) Write rules that would make this game fair.

   ....

33 ANSWERS

STARTER ACTIVITY: HOW MANY BUNNIES?

1. 6 different types of chocolate bunny are made: dark and small, dark and medium, dark and large, milk and small, milk and medium, milk and large

2. 9 different types of chocolate bunny are made – multiply 3 by 3

MAIN ACTIVITY: TWO OR MORE EVENTS

1. a) Answers through discussion. For both strawberries and raspberries, the number with no fruit is the number of plants minus the number with fruit.

b)

	Strawberries	Raspberries	Total
Fruit	20	10	30
No fruit	15	5	20
Total	35	15	50

2. a) 0.7   b) 0.2   c) 0.6

3. a) $\frac{1}{3}$ b) $\frac{2}{3}$

MAIN ACTIVITY: TREE DIAGRAMS

1. a)

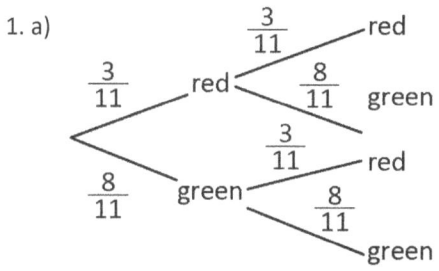

b) i) $\frac{64}{121}$ ii) $\frac{48}{121}$ iii) $\frac{112}{121}$

c) Yes. The denominator in each fraction in the second column will be 10. The numerator in the second column will change where that colour marble was chosen in the first column.

d) i) $\frac{28}{55}$ ii) $\frac{24}{55}$ iii) $\frac{52}{55}$

HOMEWORK ACTIVITY: A FAIR GAME

1. a)

		Dice 1					
		1	**2**	**3**	**4**	**5**	**6**
Dice 2	**1**	2	3	4	5	6	7
	2	3	4	5	6	7	8
	3	4	5	6	7	8	9
	4	5	6	7	8	9	10
	5	6	7	8	9	10	11
	6	7	8	9	10	11	12

b) The game is unfair because you are more likely to score 6 or more than you are to score less than 6.

c) Any game that splits the values so there are 18 ways of each player winning

d)

		Dice 1					
		1	**2**	**3**	**4**	**5**	**6**
Dice 2	**1**	1	2	3	4	5	6
	2	2	4	6	8	10	12
	3	3	6	9	12	15	18
	4	4	8	12	16	20	24
	5	5	10	15	20	25	30
	6	6	12	18	24	30	36

e) Any game that splits the values so there are 18 ways of each player winning

34 STATISTICS: PLANNING AN INVESTIGATION AND DATA COLLECTION

LEARNING OBJECTIVES

- Write down whole number values that satisfy inequalities
- Specify a problem and plan an investigation; design and use data collection sheets
- Recognise different types of data
- Infer properties of populations or distributions from a sample, while knowing the limitations of sampling

SPECIFICATION LINKS

- A22, S1

STARTER ACTIVITY

- **Which whole numbers?; 5 minutes; page 216**
  Full instructions are given on the activity sheet.

MAIN ACTIVITIES

- **Planning an investigation; 25 minutes; page 217**
  Explain to the student that you are going to find out about the exercising habits of people in the UK for a sportswear manufacturer. Discuss how they would do this, using the questions on the activity sheet to consider the following:
  a) The type of data they will be collecting: *'What sort of information will the manufacturer need? Is it primary/secondary/quantitative/qualitative?'*
  b) Fairness: To ensure that every member of the population has an equal chance of being chosen, encourage the student to think about how they can choose different ages, genders, ethnic backgrounds, social backgrounds and physical locations.
  c) Sampling: The sample needs to be large enough to be representative of the population, but not so large that the survey is too time-consuming or expensive. Discuss how to ensure the sample is random and fair (for example, by choosing people from the electoral roll, carrying out a telephone survey, stratified sampling).
  Ask the student to design a data collection sheet. Encourage them to write questions that are quick and easy to answer, and emphasise that the answer options for grouped data should not overlap or have gaps between them, and should have equal width.
- **Stratified sampling; 15 minutes; page 218**
  Work through the example together; use the language of proportion and encourage the student to recognise that the proportion of the population is equal to the proportion of the sample.

PLENARY ACTIVITY

- **The limitations of sampling; 5 minutes**
  Discuss the problems with taking a sample and discuss the limitations that even a stratified sample has (it is not necessarily a true representation of the population). You may wish to link this to the failure of the polls during the EU referendum.

HOMEWORK ACTIVITY

- **Exam-style questions; 30 minutes; page 219**
  Full instructions are given on the activity sheet.

SUPPORT IDEA

- **Planning an investigation** Show the following question to the student:
  'How many hours a week do you spend exercising: 0 to 1, 1 to 2, or 2 to 3?'
  Discuss the problems with the groups (they overlap; there is no option for more than 3 hours). Ask the student to redesign the question.

EXTENSION IDEA

- **Stratified sampling** Discuss what you should do if, when you calculate the number of people to survey in a stratum, the answer is not a whole number.

PROGRESS AND OBSERVATIONS

STARTER ACTIVITY: WHICH WHOLE NUMBERS?   TIMING: 5 MINS

LEARNING OBJECTIVES

- Write down whole number values that satisfy inequalities

EQUIPMENT

1. Write down all the whole number values that satisfy all three of these inequalities:

$-3.5 < x \leq 1$

$0.5 \leq x < \dfrac{7}{4}$

$-5.2 < x < 7$

MAIN ACTIVITY: PLANNING AN INVESTIGATION

TIMING: 25 MINS

LEARNING OBJECTIVES

- Specify a problem and plan an investigation; design and use data collection sheets
- Recognise different types of data
- Infer properties of populations or distributions from a sample, while knowing the limitations of sampling

EQUIPMENT

You are going to carry out an investigation into the sporting habits of people in the UK.

1. **Discuss how you are going to carry out this survey. Use these questions to help you.**

    a) What type of data will you be recording: primary, secondary, quantitative, qualitative?

    ....

    b) How can you ensure you carry out a fair investigation?

    ....

    ....

    ....

    c) How large should your sample be and how are you going to ensure you get a random sample?

    ....

    ....

2. **Design a data collection sheet you could use to find out about the sporting habits of people in the UK.**

    Make sure you include some questions about things like age, gender and ethnic group so you can ensure your data is not biased.

| MAIN ACTIVITY: STRATIFIED SAMPLING | TIMING: 15 MINS |

LEARNING OBJECTIVES

- Infer properties of populations or distributions from a sample, while knowing the limitations of sampling

EQUIPMENT

1. A chocolate factory makes three types of chocolate: white, milk and dark. The table shows the percentage of total sales for each type of chocolate in one year.

   Each week, the factory can produce 2000 chocolate bars. How many of these should be white chocolate?

Type of chocolate	% sales
white	12%
dark	49%
milk	39%

2. There are 500 students at a school.
   The table shows how many are in each year.

   A sample of 60 students will be taken from the whole population.
   How many students should be taken from each year?

Year group	Number of students
1	125
2	100
3	75
4	150
5	50

3. In a colony of penguins, 120 are younger than 2 years old and 180 are over 2 years old. A sample of 20 penguins is going to be taken. How many under 2 years old should be sampled?

4. The population of the UK is approximately 65 million. Approximately 33 million people are female and 32 million are male. A survey of 1000 people is planned.

   a) Write down the ratio of males to females who should be surveyed.

   b) How many males and how many females should be surveyed?

MATHS
— FOUNDATION —

AQA

HOMEWORK ACTIVITY: EXAM-STYLE QUESTIONS　　　　　**TIMING: 30 MINS**

LEARNING OBJECTIVES

- Specify a problem and plan an investigation; design and use data collection sheets
- Recognise different types of data
- Infer properties of populations or distributions from a sample, while knowing the limitations of sampling

EQUIPMENT

---

1. A scientist is recording information about the total rainfall over a one month period. She decides to contact the local meteorological office and ask for records on the rainfall. Which type of data would this be? Circle your answers.

    primary　　　　　　　secondary　　　　　　　qualitative　　　　　　　quantitative

2. A librarian records the number of books each person borrows from the library. He designs this tally chart to record the information.

Number of books	Tally
0 to 2	
2 to 4	
4 to 6	

    a) Give two problems with the tally chart.

    .... **(2 marks)**

    b) Design an improved tally chart for him to use in the space below.

    (1 mark)

3. Adrian is carrying out a survey to discover the school's favourite TV show. He decides to ask 10 people in his class what their favourite show is.

    a) Give two problems with his sample.

    .... **(2 marks)**

    b) Suggest a way he could get a random sample.

    .... **(1 mark)**

4. A gym has 1000 male and 1500 female members. The manager is carrying out a survey of 100 members of the gym. How many men should she survey?

    .... **(1 mark)**

34 Answers

STARTER ACTIVITY: WHICH WHOLE NUMBERS?

1. 1

MAIN ACTIVITY: PLANNING AN INVESTIGATION

1–2. Student's own answers

MAIN ACTIVITY: STRATIFIED SAMPLING

1. 240
2. Year 1 = 15 students; Year 2 = 12 students; Year 3 = 9 students; Year 4 = 18 students; Year 5 = 6 students
3. 8
4. a) 32 : 33   b) 492 males and 508 females

HOMEWORK ACTIVITY: EXAM-STYLE QUESTIONS

1. secondary, quantitative
2. a) The groups overlap. There is no category for someone who takes out more than 6 books.

b)

Number of books	Tally
0 – 2	
3 – 5	
6 or more	

3. a) It is not a random sample as not everyone in the school has an equal chance of being chosen. It is not a large enough sample.
b) Give each person in the school a number. Randomly generate numbers and survey those individuals.
4. 40

GLOSSARY

Primary data
Data collected by you

Secondary data
Data collected by someone else and used by you

Quantitative data
Data that has a numerical value

Qualitative data
Data that is non-numerical

Stratified sample
A sample that contains members of each group in proportion to the sizes of each group in relation to the whole population

35 STATISTICS: CONSTRUCTING GRAPHS, CHARTS AND DIAGRAMS

LEARNING OBJECTIVES

- Calculate fractions or percentages of whole numbers
- Interpret and construct tables, charts and diagrams, including frequency tables, bar charts and pie charts, for categorical data and ungrouped discrete numerical data

SPECIFICATION LINKS

- S2

STARTER ACTIVITY

- **360; 5 minutes; page 222**
  Ask the student to calculate the fractions and percentages of 360 shown on the activity sheet.

MAIN ACTIVITIES

- **Displaying data; 25 minutes; page 223**
  Look at the stem-and-leaf diagram with the student and ensure they understand it by asking questions like '*How old is the oldest female patient? The youngest female patient? How many female patients are there?*' Ask the student to complete the back-to-back stem-and-leaf diagram using the data on male patients. When the student transfers the data to the frequency table (question 2), discuss the loss of accuracy of the data, but establish that it makes it simpler to use for charts and diagrams. Then ask the student to complete the remaining questions.
- **Comparing pie charts; 15 minutes; page 224**
  Look at the two pie charts shown and discuss what each sector represents. Model some calculations, showing the student how to work out how many students one of the sectors represents. Then ask the student to work out how many students one of the other sectors represents.
  Work through the 'true/false' questions, encouraging the student to justify their answers using calculations as well as explaining in words. The student is then challenged to write their own true statements about the information shown.

PLENARY ACTIVITY

- **Which is best?; 5 minutes**
  Ask the student to give the advantages and disadvantages of using each of the types of diagram, chart and graph listed below to display data. Is it particularly appropriate for a certain kind of data? Ways of displaying data are:
  pie chart, bar chart, comparative bar chart, dual bar chart, stem-and-leaf diagram, frequency table

HOMEWORK ACTIVITY

- **Media search; 40 minutes; page 225**
  Full instructions are given on the activity sheet.

SUPPORT IDEAS

- **Displaying data** For the frequency table, provide the groupings: $0 < x \leq 20$, $20 < x \leq 40$, $40 < x \leq 60$, $60 < x \leq 80$, $80 < x \leq 100$.
- **Media search** You may wish to source some graphs, charts and diagrams for the student to discuss.

EXTENSION IDEA

- **Comparing pie charts** Challenge the student to draw a comparative bar chart to represent the data shown in both pie charts.

PROGRESS AND OBSERVATIONS

STARTER ACTIVITY: 360

TIMING: 5 MINS

LEARNING OBJECTIVES

- Calculate fractions or percentages of whole numbers

EQUIPMENT

1. **Work out these fractions and percentages of 360.**

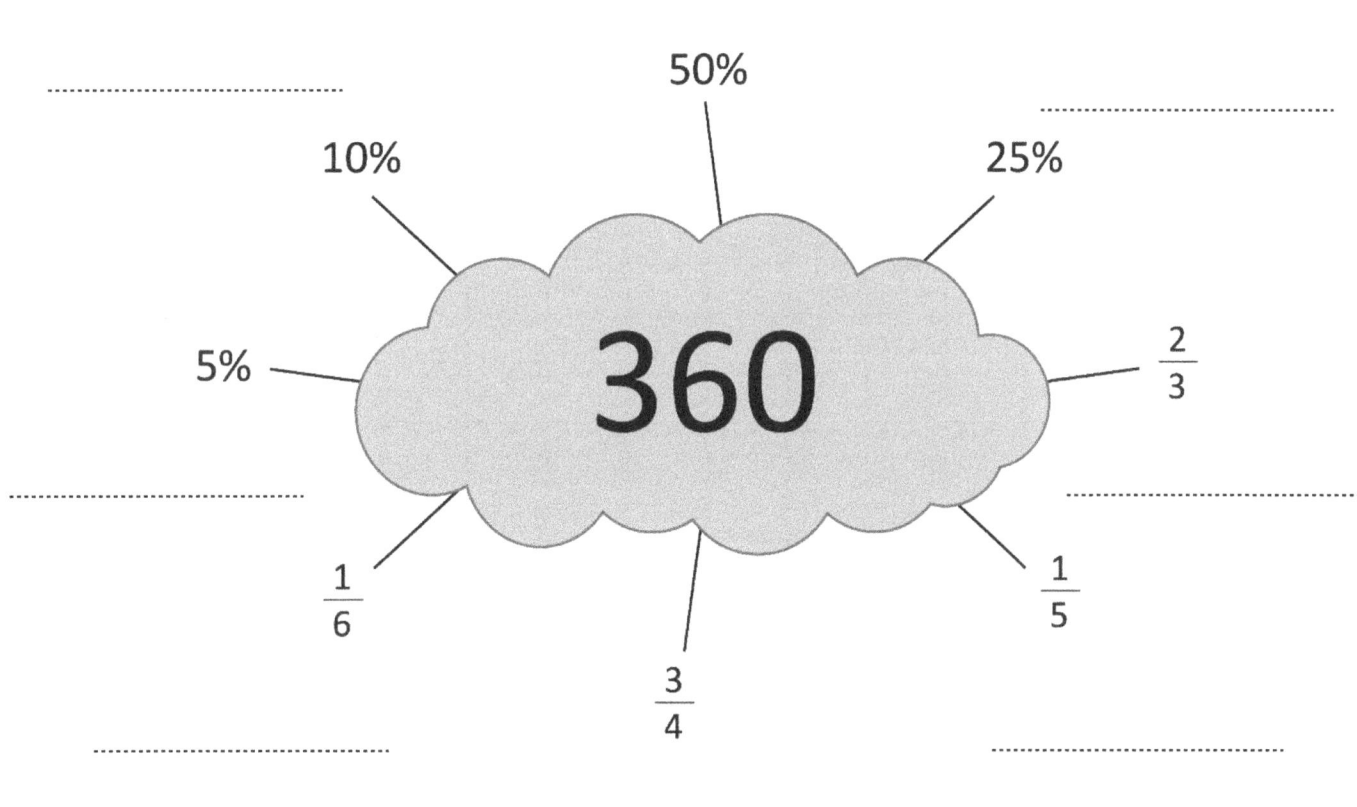

MAIN ACTIVITY: DISPLAYING DATA

TIMING: 25 MINS

LEARNING OBJECTIVES

- Interpret and construct tables, charts and diagrams, including frequency tables, bar charts and pie charts, for categorical data and ungrouped discrete numerical data

EQUIPMENT

- ruler
- squared paper

1. **The partially-completed stem-and-leaf diagram shows the ages of the male and female patients in a hospital.**

Male		Female
	1	8 9 9 9
	2	3 4 6
	3	0 9
	4	2 3 3 5 6
	5	3 4 8
	6	0 4 5 9 9
	7	0 0 7 7 8 9
	8	2 5 6 6
	9	0 1

Key
Female: 1|8 means 18 years
Male: 8|1 means 18 years

The ages of the male patients are:
88, 41, 25, 57, 50, 80, 22, 57, 20, 64, 19, 66, 28, 55, 62, 87, 83, 63, 87, 75, 83, 69, 56, 27, 66, 43

 a) Complete the back-to-back stem-and-leaf diagram.

 b) Display all the data in a grouped frequency table. Aim for four or five groups of equal width. Check you have included all the data by totalling the frequency. There are 60 patients in total.

 c) Display the data from the grouped frequency table in either a bar chart or a pictogram.

 d) How could a dual or compound bar chart could be used to display the data. Why might this be a good way of displaying the original data?

MAIN ACTIVITY: COMPARING PIE CHARTS

TIMING: 15 MINS

LEARNING OBJECTIVES

- Interpret pie charts for categorical data

EQUIPMENT

Two schools record the ways
their students travel to school.

The information is represented in
these two pie charts.

Southdean School has 2000 students.
Northholt College has 500 students.

Southdean School

bus 40%   walk 45%   train 5%   bike 10%

Northolt College

bus 40%   walk 55%   bike 5%

1. Decide if the following statements are true or false. Give reasons for your answers.

   a) The number of students who take the bus to school is the same at both schools.

   b) The proportion of students who take the train to Southdean is equal to the proportion who ride a bike to Northholt.

   c) More students walk to Northholt than walk to Southdean.

   d) Fewer students catch the train to Northolt than catch the train to Southdean.

   e) The fraction of students who ride a bike to Northolt is smaller than the fraction who ride a bike to Southdean.

2. Write down three other true statements about the transport of the students at the two schools.

HOMEWORK ACTIVITY: MEDIA SEARCH

TIMING: 40 MINS

LEARNING OBJECTIVES

- Interpret and construct tables, charts and diagrams, including frequency tables, bar charts and pie charts for categorical data and ungrouped discrete numerical data

EQUIPMENT

1. **Graphs, charts and tables are used all the time to display data in the media.**

   Find at least three different examples of graphs, charts and/or tables in newspapers, magazines or on the internet.

   For each example you find:

   a) decide whether the graph, chart or table has been drawn accurately. (For example, if the data is grouped, do the groups all have equal width?)

   b) decide whether you think this is the best way to display the data.

   c) explain any ways in which you think the graph, chart or table is misleading.

35 Answers

STARTER ACTIVITY: 360

1. 5% = 18, 10% = 36, 50% = 180, 25% = 90, $\frac{1}{6}$ = 60, $\frac{3}{4}$ = 270, $\frac{1}{5}$ = 72, $\frac{2}{3}$ = 240

MAIN ACTIVITY: DISPLAYING DATA

1. a)

Male		Female
9	1	8 9 9 9
8 7 5 2 0	2	3 4 6
	3	0 9
3 1	4	2 3 3 5 6
7 7 6 5 0	5	3 4 8
9 6 6 4 3 2	6	0 4 5 9 9
5	7	0 0 7 7 8 9
8 7 7 3 3 0	8	2 5 6 6
	9	0 1

b)

Age (years)	Frequency
$0 < A \le 20$	6
$20 < A \le 40$	9
$40 < A \le 60$	16
$60 < A \le 80$	18
$80 < A \le 100$	11

c) Check student's bar chart or pictogram.

d) A dual or compound bar chart could have separate bars or sections for males and females. This would make it easier to compare the number of males and females within each age group.

MAIN ACTIVITY: COMPARING PIE CHARTS

1. a) False – 40% of students at both schools take the bus, but the total number of students at each school is different, so the number of students taking the bus is different: 40% of 2000 = 800 Southdean students take the bus; 40% of 500 = 200 Northholt students take the bus.

b) True – 5% of Southdean students take the train and 5% of Northholt students ride a bike.

c) False – A higher proportion of Northholt students walk (55% compared with 45%), but a smaller number of Northholt students walk: 55% of 500 = 275 Northholt students walk; 45% of 2000 = 900 Southdean students walk.

d) True – 5% of Southdean students take the train, and no Northholt students take the train.

e) True – 5% = $\frac{1}{20}$ of Northholt students ride a bike; 10% = $\frac{1}{10}$ of Southdean students ride a bike.

2. Student's own answers

HOMEWORK ACTIVITY: MEDIA SEARCH

1. Student's own work

36 STATISTICS: MEASURES OF SPREAD AND LOCATION

LEARNING OBJECTIVES

- Interpret, analyse and compare distributions of data sets from univariate empirical distributions through appropriate measures of central tendency (median, mean, mode and modal class) and spread (range)
- Apply statistics to describe a population

SPECIFICATION LINKS

- S4, S5

STARTER ACTIVITY

- **True/false/can't tell; 5 minutes; page 228**
  Full instructions are given on the activity sheet.

MAIN ACTIVITIES

- **Mean, median, mode and range; 20 minutes; page 229**
  Establish how to calculate the mean, median, mode and range of a simple set of numbers. In the activity, two sets of scores are given for two contestants in a dance competition. Ask the student to calculate the mean, mode, median and range of each set of data. Discuss which dancer they think is better by considering the averages and spread. Emphasise that as long as they justify their answer and back it up with mathematical calculations, they could choose either.
- **Working with averages and spread; 20 minutes; page 230**
  The task requires the generation of random numbers, which can be done using a calculator or by picking cards from a shuffled set of digit cards. If you do not have suitable equipment available, you could choose the numbers yourself. Encourage the student to think carefully about the questions on the activity sheet and to explain their thought processes as they work through the problems.

PLENARY ACTIVITY

- **Ten words only; 5 minutes**
  Ask the student to explain the terms mean, mode, median and range using no more than ten words for each description.

HOMEWORK ACTIVITY

- **Tick or trash?; 30 minutes; page 231**
  Full instructions are given on the activity sheet.

SUPPORT IDEA

- **Working with averages and spread** For parts a) to f), ask the student to calculate the new mean, mode, median and range first, and then to justify any changes to these values.

EXTENSION IDEA

- **Working with averages and spread** Randomly generate five numbers and do not show them to the student. Work out the mean, mode, median and range and ask them to suggest what the five numbers could be.

PROGRESS AND OBSERVATIONS

STARTER ACTIVITY: TRUE/FALSE/CAN'T TELL

TIMING: 5 MINS

LEARNING OBJECTIVES

- Interpret frequency tables

EQUIPMENT

A group of people were asked how many siblings they had. The results were recorded in the table.

Number of siblings	Frequency
0	3
1	2
2	7
3	1
4+	2

1. Decide whether each of these statements is true or false, or whether you can't tell from the table.

	true	false	can't tell
a) More people had three siblings than two siblings.	☐	☐	☐
b) The highest number of siblings of anyone interviewed was five.	☐	☐	☐
c) There were three people with three or more siblings.	☐	☐	☐
d) Five people had two siblings or fewer.	☐	☐	☐
e) The total number of siblings the group had can be calculated by working out 3 + 2 + 7 + 1 + 2.	☐	☐	☐

MAIN ACTIVITY: MEAN, MEDIAN, MODE AND RANGE TIMING: 20 MINS

LEARNING OBJECTIVES

- Interpret, analyse and compare distributions of data sets from univariate empirical distributions through appropriate measures of central tendency (median, mean, mode and modal class) and spread (range)

EQUIPMENT

1. **In a dance competition, the competitors were marked out of 10. Their scores are shown in this table.**

Contestant A	9	4	8	8	7	8	2	8	8
Contestant B	7	8	8	7	7	9	6	7	8

a) Work out the mode, median, mean and range for each contestant.

Contestant A:

mode = ...

median = ...

mean = ...

range = ...

Contestant B:

mode = ...

median = ...

mean = ...

range = ...

b) Which dancer would you give the prize to? Explain your answer.

....

....

MAIN ACTIVITY: WORKING WITH AVERAGES AND SPREAD   TIMING: 20 MINS

LEARNING OBJECTIVES

- Interpret, analyse and compare distributions of data sets from univariate empirical distributions through appropriate measures of central tendency (median, mean, mode and modal class) and spread (range)

EQUIPMENT

- random number generator

Randomly generate five 1-digit numbers and write them on the lines.

....

1. For the five numbers, calculate:

    a) the mean   ....   b) the median   ....

    c) the mode   ....   d) the range.   ....

2. Explain what would happen to the mean, mode, median and range if:

    a) you added 10 to the largest number

    ----

    b) you added 10 to the smallest number

    ----

    c) you added 10 to the middle number (if they were written in order)

    ----

    d) you included a sixth number which was equal to the largest number

    ----

    e) you included a sixth number which was larger than the largest number

    ----

    f) you included a sixth number which was equal to the smallest number.

HOMEWORK ACTIVITY: TICK OR TRASH?

TIMING: 30 MINS

LEARNING OBJECTIVES

- Interpret, analyse and compare distributions of data sets from univariate empirical distributions through appropriate measures of central tendency (median, mean, mode and modal class) and spread (range)
- Apply statistics to describe a population

EQUIPMENT

 1. Ahmed's results in his GCSEs are: 9 5 7 8 6 6 5 4 7 6

Janice's results are: 9 9 6 3 8 7 5 7 9 2

For each of the statements below, decide whether to tick it (if it is correct) or trash it (if it is false). Circle the correct icon to show your decision.

	Tick or trash?	
a) Ahmed's mean is higher than his mode.	✓	🗑
b) Janice's median is higher than her mode.	✓	🗑
c) Ahmed's mean is higher than Janice's mode.	✓	🗑
d) If Ahmed had scored 2 higher on his lowest mark, he would have had the same mean as Janice.	✓	🗑
e) Ahmed's range is lower than Janice's.	✓	🗑
f) Janice's mean is higher than Ahmed's.	✓	🗑
g) Ahmed's median is equal to his mode.	✓	🗑
h) Janice's mean is equal to her median.	✓	🗑
i) Janice and Ahmed have the same median score.	✓	🗑
j) The sum of Janice and Ahmed's means is 12.8.	✓	🗑

2. Amelie sat 10 GCSEs. Her mode was 3, her mean was 3 and her median was 3. Her range was 5. Give one possible set of marks for Amelie.

36 Answers

STARTER ACTIVITY: TRUE/FALSE/CAN'T TELL

1. a) false   b) can't tell   c) true   d) false   e) false

MAIN ACTIVITY: MEAN, MEDIAN, MODE AND RANGE

1. a) Contestant A: mode = 8   median = 8   mean = 6.9 (1 decimal place)   range = 7
Contestant B: mode = 7   median = 7   mean = 7.4 (1 decimal place)   range = 3
b) Either is fine, as long as the student justifies their answer with reference to the averages and spread.

MAIN ACTIVITY: WORKING WITH AVERAGES AND SPREAD

1. Check student's answers.

2. a) The mean would increase by 2; if the mode was originally the largest number it might change, otherwise it would remain unchanged; the median would not change; the range would increase by 10.

b) The mean would increase by 2; if the mode was originally the smallest number it might change, otherwise it would remain unchanged; the median might increase; the range would change.

c) The mean would increase by 2; if the mode was originally the middle number it might change, otherwise it would remain unchanged; the median might increase; the range might increase.

d) The mean would increase; the mode might change to the largest number; the median might increase; the range would remain unchanged.

e) The mean would increase; the mode would remain unchanged; the median might increase; the range would increase.

f) The mean would decrease; the mode might change to the smallest number; the median might decrease; the range would remain unchanged.

HOMEWORK ACTIVITY: TICK OR TRASH

1. Ahmed: mean = 6.3   median = 6   mode = 6   range = 5
Janice: mean = 6.5   median = 7   mode = 9   range = 7
a) tick   b) trash   c) trash   d) tick   e) tick
f) tick   g) tick   h) trash   i) trash   j) tick
2. Student's own answer

GLOSSARY

Mean
An average found by adding together all the values in a data set and then dividing by the number of values

Mode
The value in a data set that occurs most often

Median
The middle value when the values in a data set are written in order from smallest to largest

Range
The difference between the largest value and the smallest value in a data set

37 STATISTICS: USING MEASURES OF LOCATION AND SPREAD

LEARNING OBJECTIVES

- Find the midpoint between two whole number values
- Apply statistics to describe a population
- Interpret tables, charts and diagrams
- Interpret, analyse and compare the distributions of data sets from univariate empirical distributions through: appropriate graphical representations involving discrete, continuous and grouped data; appropriate measures of central tendency (median, mean, mode and modal class) and spread (range, including consideration of outliers)

SPECIFICATION LINKS

- S2, S4

STARTER ACTIVITIES

- **Midpoints; 5 minutes; page 234**
  Full instructions are given on the activity sheet.

MAIN ACTIVITY

- **Calculating statistics from tables and graphs; 30 minutes; page 235**
  Remind the student how to find the mean, mode and median of a set of data. Work through the activities, finding the measure of location and spread listed. When working with grouped data, explain that the midpoint is used. Establish that this is why the mean is an estimate (since the exact data values are not given).

- **Using averages and spread; 10 minutes; page 236**
  Explain what an outlier is. Establish that although sometimes you can discount outliers, you may not be able to. Discuss with the student the effects that an outlier has on the different measures of spread and location. Work through the activity sheet, encouraging the student to explain their reasoning.

PLENARY ACTIVITY

- **Which numbers?; 5 minutes**
  Explain that you are thinking of three numbers: the mean of the numbers is 12 and the mode is 15. What is the range of the numbers? (The numbers are 6, 15 and 15 so range is 9.) Challenge the student to design a question like this for you.

HOMEWORK ACTIVITY

- **Revision cards; 45 minutes; page 237**
  Full instructions are given on the activity sheet.

SUPPORT IDEAS

- **Calculating statistics from tables and graphs** When working with the frequency tables, you may wish to ask the student to start by writing out a list of the data values, e.g. 2, 2, 2, 2, 3, ... etc.
- **Using averages and spread** Give the student a set of data to experiment with. What happens to the measures of location and spread when you add a very high/low value?

EXTENSION IDEA

- **Using averages and spread** Encourage the student to work out all possible average times and discuss the benefits of each one in different situations. Which one will make the car sound the best? Which one is the most honest?

PROGRESS AND OBSERVATIONS

STARTER ACTIVITY: MIDPOINTS   TIMING: 5 MINS

LEARNING OBJECTIVES	EQUIPMENT
• Find the midpoint between two whole number values	none

We can work out which number is halfway between a pair of numbers by adding them together and dividing by 2.

Example:
Find the number halfway between 12 and 23.

$$\frac{12+23}{2} = 17.5$$

---

1. **Work out which number is halfway between:**

    a) 10 and 35 ....

    b) 0 and 17 ....

    c) −8 and 4 ....

    d) *a* and *b*. ....

MAIN ACTIVITY: CALCULATING STATISTICS FROM TABLES AND GRAPHS   TIMING: 30 MINS

LEARNING OBJECTIVES

- Interpret tables, charts and diagrams
- Interpret, analyse and compare the distributions of data sets from univariate empirical distributions through: appropriate graphical representations involving discrete, continuous and grouped data; appropriate measures of central tendency (median, mean, mode and modal class) and spread (range, including consideration of outliers)

EQUIPMENT

- calculator

For each of the tables, graphs and charts, calculate the averages and/or range as specified.

1. **This table shows the number of sports played by 50 gym users.**

   mean = ...   mode = ...

   median = ...   range = ...

Number of sports played	Frequency
2	5
3	38
4	6
5	1

2. **This bar chart shows the number of visits made to the gym per week by 50 customers.**

   mean = ...

   mode = ...

   range = ...

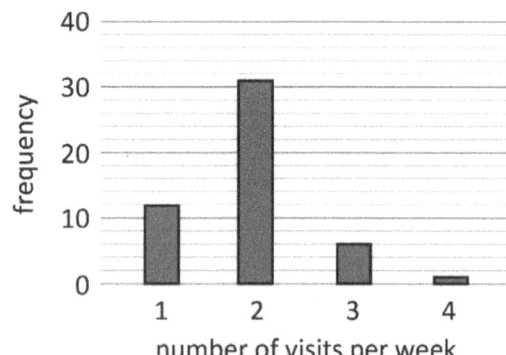

3. **This table shows the age of customers in a gym.**

   mean = ...

   modal class = ...

   class containing the median = ...

Age (x)	Frequency
$0 < x \le 20$	3
$20 < x \le 40$	14
$40 < x \le 60$	11
$60 < x \le 80$	2

4. **This box plot shows the number of minutes customers spent in the gym one morning.**

   median = ...

   range = ...

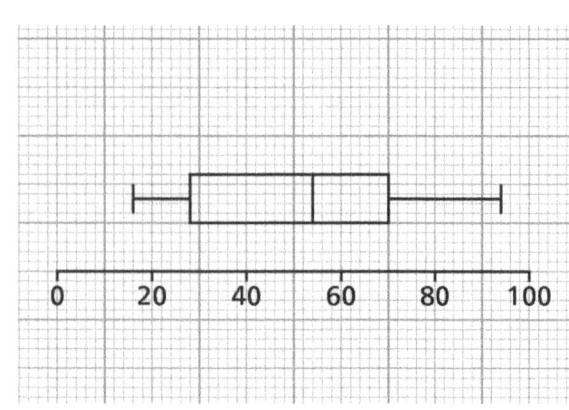

MATHS
— FOUNDATION —

MAIN ACTIVITY: USING AVERAGES AND SPREAD

TIMING: 10 MINS

LEARNING OBJECTIVES

- Know the advantages and disadvantages of the measures of central tendency (median, mean, mode and modal class) and spread (range, including consideration of outliers)

EQUIPMENT

1. Decide if each of these statistics will be affected or unaffected by an outlier. Explain your answers.

   a) mode

   b) median

   c) mean

   d) range

2. A car manufacturer records the time in seconds it takes a car to accelerate from 0 to 60 miles per hour on ten different occasions. The times are: 5.6, 5.7, 6.0, 5.8, 5.6, 5.7, 5.9, 5.9, 5.6, 10.2.

   Which average should the car manufacturer use in its adverts? Explain your reasoning.

HOMEWORK ACTIVITY: REVISION CARDS

TIMING: 45 MINS

LEARNING OBJECTIVES

- Interpret, analyse and compare the distributions of data sets from univariate empirical distributions through: appropriate graphical representations involving discrete, continuous and grouped data; appropriate measures of central tendency (median, mean, mode and modal class) and spread (range, including consideration of outliers)

EQUIPMENT

- index cards

1. **Make a set of four revision cards, one on each of the following topics:**

   - mean

   - median

   - mode

   - range.

   Make sure you include information on:

   - how to calculate them

   - the advantages and disadvantages of each

   - how you can calculate them from at least one graph, table or chart.

37 ANSWERS

STARTER ACTIVITY: MIDPOINTS

1. a) 22.5   b) 8.5   c) −2   d) $\frac{a+b}{2}$

MAIN ACTIVITY: CALCULATING STATISTICS FROM TABLES AND GRAPHS

1. mean = 3.06	mode = 3	median = 3	range = 3
2. mean = 1.92	mode = 2	range = 3	
3. mean = 38	modal class = 20 < x ≤ 40	class containing the median = 20 < x ≤ 40	
4. median = 54	range = 78		

MAIN ACTIVITY: USING AVERAGES AND SPREAD

1. a) unaffected	b) unaffected	c) affected	d) affected

2. mean = 6.2, mode = 5.6, median = 5.75

The manufacturer should use the mode as it is the smallest value. The mean is affected by the outlier.

HOMEWORK ACTIVITY: REVISION CARDS

1. Student's own work

GLOSSARY

Outlier

A data value which does not fit the pattern of the data. It is much smaller or larger than the other values.

38 STATISTICS: SCATTER GRAPHS

LEARNING OBJECTIVES

- Use and interpret scatter graphs of bivariate data; recognise correlation and know that it does not indicate causation; draw estimated lines of best fit; make predictions; interpolate and extrapolate apparent trends while knowing the dangers of doing so

SPECIFICATION LINKS

- S6, A14

STARTER ACTIVITY

- **Conversion graph; 5 minutes; page 240**
  Show the student the graph and discuss how it can be used. Explain that it shows the conversion rate between British pounds and American dollars (in February 2017). Remind the student that a graph like this is a pictorial way of representing a relationship between two variables. Ask the student to convert varied amounts in £s into $s and vice-versa.

MAIN ACTIVITIES

- **Plotting scatter graphs; 20 minutes; page 241**
  Discuss with the student when a scatter diagram might be useful (when you are showing a relationship between two sets of values). Look at the data given on the activity sheet and discuss how to plot this on a scatter diagram. Ensure that the student draws and labels the scales and axes accurately. Ask the student to plot the points of the scatter diagram, marking each point with a cross. Establish how to draw a line of best fit and ask the student to read some values from the line of best fit. Discuss the accuracy of these results.

- **Correlation, outliers, extrapolation; 20 minutes; page 242**
  Full instructions are given on the activity sheet. Spend time discussing the difference between correlation and causation, giving the example of the positive correlation between consumption of chocolate per capita and Nobel prize winners per capita to show that two things in correlation are not always cause and effect.

PLENARY ACTIVITY

- **The language of scatter graphs; 5 minutes**
  Ask the student to explain all the terms they have used in this session that are associated with scatter graphs.

HOMEWORK ACTIVITY

- **All about scatter graphs; 60 minutes; page 243**
  Full instructions are given on the activity sheet.

SUPPORT IDEA

- **Plotting scatter graphs** Provide the axes for the student with intervals already marked and labelled.

EXTENSION IDEA

- **Correlation, outliers and extrapolation** Challenge the student to give three instances where correlation implies causation and three instances where it does not.

PROGRESS AND OBSERVATIONS

STARTER ACTIVITY: CONVERSION GRAPH

TIMING: 5 MINS

LEARNING OBJECTIVES

• Use and interpret scatter graphs of bivariate data

EQUIPMENT

1. This graph shows the conversion rate between British pounds and American dollars (in February 2017). Answer the questions your tutor asks.

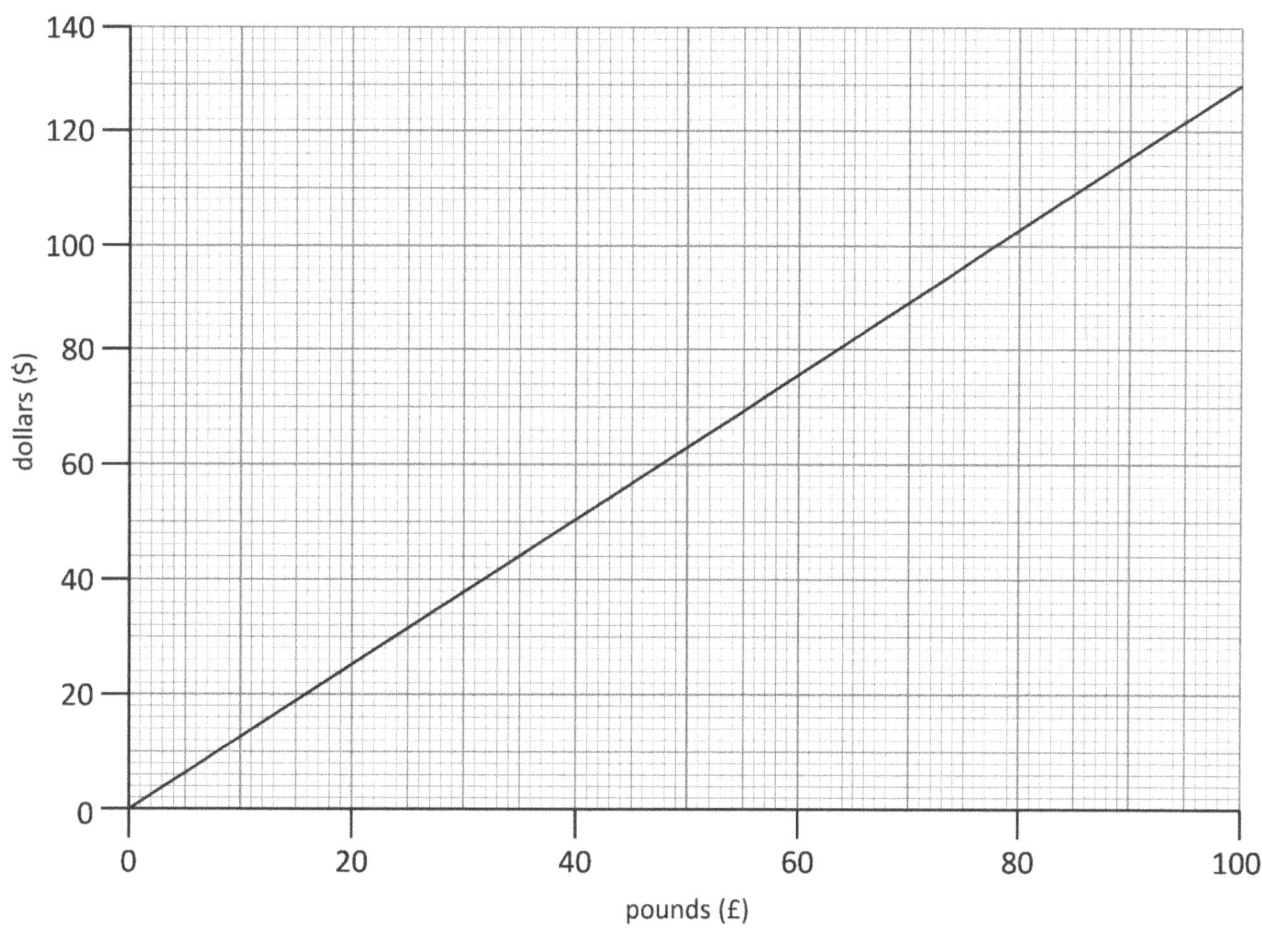

MATHS
— FOUNDATION —

AQA

MAIN ACTIVITY: PLOTTING SCATTER GRAPHS   **TIMING: 20 MINS**

LEARNING OBJECTIVES

- Use and interpret scatter graphs of bivariate data; draw estimated lines of best fit

EQUIPMENT

- ruler

1. The weights and lengths of 10 new-born babies are recorded.

Weight (kg)	3.5	3.7	3.9	3.1	4.2	3.8	3.9	3.1	4.4	3.7
Length (cm)	50.3	52	54.1	48.5	53.1	54.2	52.1	45.2	56	50.5

a) Draw a scatter diagram to show this data.

b) Draw a line of best fit on the graph. Explain what this line shows.

....

MAIN ACTIVITY: CORRELATION, OUTLIERS, EXTRAPOLATION   TIMING: **20** MINS

LEARNING OBJECTIVES	EQUIPMENT
• Use and interpret scatter graphs of bivariate data; recognise correlation and know that it does not indicate causation; draw estimated lines of best fit; make predictions; interpolate and extrapolate apparent trends while knowing the dangers of doing so	none

1. **For each graph, decide if the relationship between *x* and *y* shows positive, negative, or no correlation.**

a)   b)   c)   d)

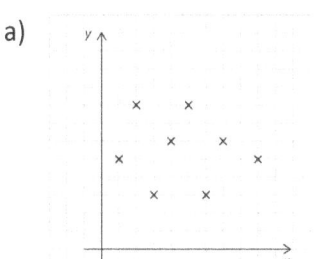 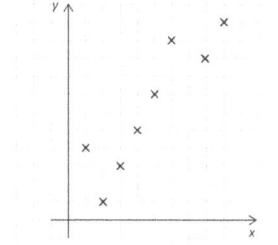 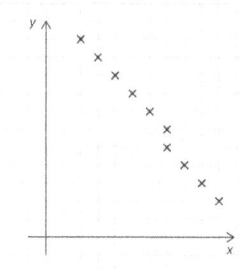

....

2. **Look at the graph you drew in the previous activity. What type of correlation does this graph show?**

----

3. **A reporter investigated the price of a three bedroom house and the length of time it took to travel by train from the house to central London.**
   She drew a scatter diagram to show her results.

   a) What type of correlation is shown?

   ....

   b) Discuss with your tutor whether the graph shows a casual relationship.

   c) Draw a line of best fit on the diagram.

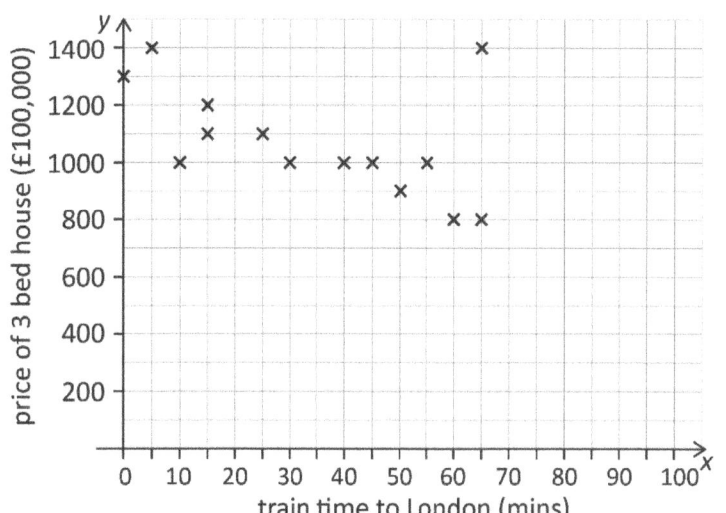

   d) Use your line of best fit to estimate the cost of a three bedroom house from which it takes 100 minutes to travel to London. How reliable is this? Explain your reasoning.

----

   e) Circle any points you consider to be outliers. Can you explain why they might not fit the rest of the data?

HOMEWORK ACTIVITY: ALL ABOUT SCATTER GRAPHS

TIMING: 60 MINS

LEARNING OBJECTIVES

- Use and interpret scatter graphs of bivariate data; recognise correlation and know that it does not indicate causation; draw estimated lines of best fit; make predictions; interpolate and extrapolate apparent trends while knowing the dangers of doing so

EQUIPMENT

- filming equipment (optional)

1. **Imagine you are a Year 11 teacher. Your task is to teach a lesson about scatter graphs. You could record a video or write a lesson plan.**

   Make sure you explain:

   - how to construct a scatter diagram

   - when you might use a scatter diagram

   - correlation

   - lines of best fit

   - extrapolation

   - outliers.

38 ANSWERS

STARTER ACTIVITY: CONVERSION GRAPH

1. Answers depend on tutor's questions.

MAIN ACTIVITY: PLOTTING SCATTER GRAPHS

1. a)

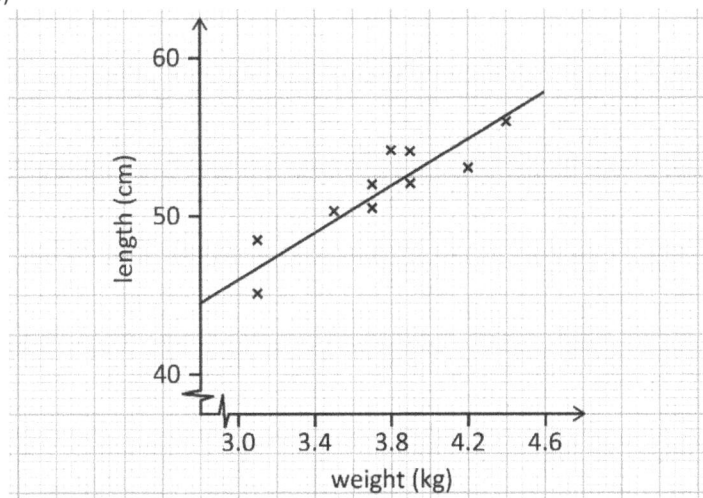

b) The line of best fit shows that as the babies' length increases, their weight also increases. There is a positive correlation between the two variables.

MAIN ACTIVITY: CORRELATION, OUTLIERS, EXTRAPOLATION

1. a) no correlation   b) negative correlation   c) positive correlation   d) negative correlation
2. positive
3. a) negative
b) You cannot tell whether there is a causal relationship. It is possible that proximity to London affects house prices, but this may not be the only factor and you cannot tell from the graph.
c) Check student's line of best fit.
d) Check student's estimate. The answer is unreliable because it is beyond the limits of the original data.
e) There is an outlier at (65, 1.4 million). This might be a house with extra features, or in a particularly exclusive area. Accept any sensible reason.

HOMEWORK ACTIVITY: ALL ABOUT SCATTER GRAPHS

1. Student's own work

GLOSSARY

Correlation
A relationship between two sets of values. Values correlate if they fall close to a straight line when plotted on a scatter graph.

Extrapolation
Predicting a value that falls outside the range of the data

PROGRESS AND OBSERVATIONS

PROGRESS AND OBSERVATIONS

Published by Pearson Education Limited, 80 Strand, London, WC2R 0RL.

www.pearsonschools.co.uk

Text © Pearson Education Limited 2017
Series consultant: Margaret Reeve
Edited by Elektra Media Ltd
Designed by Andrew Magee
Typeset by Elektra Media Ltd
Produced by Elektra Media Ltd
Original illustrations © Pearson Education Limited 2017
Illustrated by Elektra Media Ltd
Cover design by Andrew Magee
Printed in the UK by Ashford Press Ltd

The right of Catherine Murphy to be identified as author of this work has been asserted by her in accordance with the Copyright, Designs and Patents Act 1988.

First published 2017

20 19 18 17
10 9 8 7 6 5 4 3 2 1

British Library Cataloguing in Publication Data
A catalogue record for this book is available from the British Library

ISBN 978-1-292-19554-4

The ActiveBook accompanying this book contains editable Word files. Pearson Education Limited cannot accept responsibility for the quality, accuracy or fitness for purpose of the materials contained in the Word files once edited. To revert to the original Word files, download the files again.

Printed in the United Kingdom by Ashford Press Ltd

Acknowledgements

We would like to thank Tutora for its invaluable help in the development and trialling of this course.

Notes from the publisher
Pearson has robust editorial processes, including answer and fact checks, to ensure the accuracy of the content in this publication, and every effort is made to ensure this publication is free of errors. We are, however, only human, and occasionally errors do occur. Pearson is not liable for any misunderstandings that arise as a result of errors in this publication, but it is our priority to ensure that the content is accurate. If you spot an error, please do contact us at resourcescorrections@pearson.com so we can make sure it is corrected.